TH5606
M22   3077
Carpentry framing and
finishing

DATE DUE

	NOV 1 6 1988	
JUN 6 1986		
FEB 6 1987		
MAR 6 1987		
MAR 1 7 1988		
JUN 1 5 1988		
DEC 2 9 1989		
OCT 5 1989		
APR 2 1990		
DEC 1 7 1990		
JUN 4 1991		
MAY 1 6 1996		

HIGHSMITH 45-220

CARPENTRY
Framing & Finishing

CARPENTRY
Framing & Finishing

Byron W. Maguire

RESTON PUBLISHING COMPANY, INC.
A Prentice-Hall Company
Reston, Virginia

Illustrations by Melvin Blankenship

Library of Congress Cataloging in Publication Data
Maguire, Byron W
    Carpentry: framing and finishing.

    Bibliography: p.
    Includes index.
    1. Carpentry.   2. House construction.   I. Title.
TH5606.M22   694   78-31970
ISBN 0-8359-0701-5

© 1979 by Reston Publishing Company, Inc.
A Prentice-Hall Company
Reston, Virginia 22090

1   3   5   7   9   10   8   6   4   2

Printed in the United States of America

Table of Contents

Preface

"How do you know what needs to be done and the way to do it?" the apprentice asked the master carpenter. He replied, "I start with a look at the plans and then I translate what they tell me into practical tasks that produce soundly built structures. You will soon learn from experience, and by reading and studying that forces such as wind, snow, and live and dead weights are always acting on structures and can cause much damage. You will also learn that certain tasks should be done first. Studying the principles of carpentry helps you understand why I do what I do."

"I guess I had better really begin studying so I can find out what to do without always asking," the apprentice said—more to himself than to the master carpenter.

*   *   *

Beginners and apprentices are able to do many varied carpentry tasks and, for the most part, to perform adequately with a minimum of understanding. Unfortunately, many experienced carpenters who consider themselves journeymen are really only expert at those tasks they have done repeatedly, but still do not understand why.

Explanations at the engineering level would tend to confuse or discourage. For that reason this book is written as a "bridge" from the doing, to the knowing why—but not at the engineering level. This book is for apprentices and journeyman carpenters who need to know more of the "whys" of carpentry.

To begin, the master carpenter sees a structure as a group of integrated members made from all manner of materials. Studying Chapters 1 and 2 will help the reader learn to see in the same way. For example, the master carpenter looks at a window and sees the details that go into its framing, or he looks at a door and visualizes its framing. He sees from the overall design where roof joins wall and where load and stress balance.

The master carpenter also knows that all materials are not alike, and so they must be carefully chosen from the many available species and grades. Chapter 3 identifies these and the later chapters show how to make use of them. Strength and stiffness are two main characteristics of lumber. The technical terms for them are "fiber-in-bending" and "modulus of elasticity." For instance, the strength and stiffness of plywood are precisely rated because plywood is much used in construction. In addition, the current need to conserve fossil fuels is prompting the construction industry to better insulate both new and old buildings. So, in Chapter 4 the R-factors (heat gain and heat loss values) are examined. Improvements in thermal protection are thoroughly detailed.

Chapters 5 through 11 examine the principles involved in constructing all the framing for a residential building. The technical aspects are first defined then shown in examples.

Chapters 12 through 16 examine some of the more critical and exacting tasks that the journeyman and master carpenter must perform; for example, the principles and techniques involved in building a porch column or a screen enclosure correctly. Stairs and stair building, among the most complicated tasks in carpentry, are discussed in Chapters 14, 15, and 16.

These final chapters provide numerous examples of skilled carpentry. To build well, every step in each process is carefully and fully described. When the reader understands the "why" of carpentry and can construct the different assemblies that this book examines, he is well on the way to becoming a craftsman.

Many associations and suppliers of the building industries have contributed to this book. Special thanks are due the following: American Plywood Association, Boise Cascade Corporation, Celetex Corp., National Forest Products Association, Southern Forest Products Association, W. R. Grace Co., and Western Woods Products Association.

CARPENTRY
Framing & Finishing

Functional Properties in House Construction

The novice carpenter is most often thrown without training into the construction of houses by his employer, and so he takes the instructions on each task and applies them to that task. In this environment he has little if any time to reflect on why the task is done in a certain way or to notice that a building principle is involved. When the occasion arises for a repetition of the same task, he repeats it, as often as necessary. Naturally, if a variety of tasks are performed over and over, a definite skill is obtained. But, has any learning of construction principles taken place? All too often the answer is no, because when this "trained" carpenter is confronted with a slight variation of the principle applied, he cannot cope, and thus makes mistakes that result in a weakness in the structure of the house.

To be effective in applying the principles advanced in this book, the functional properties of house construction need to be defined to set a common ground for meaningful study.

ESTABLISH PARAMETERS

The first step in establishing the parameters of a house is to obtain a total concept of the exterior proportions of the house to be built. From what data can such an idea be developed, and what should be included? It could be that the concept to be formed is the *character* of the house's exterior developed from its parameters. Parameters include the shape and kind of roof or cornice, proportions of the house using straight lengths of walls and offsets, doors of common and special character, window placement as well as kind of window, and exterior wall materials. This is an ideal way of approaching the problem. If there is an architect's rendering of the building, much of its character can be gleaned from careful study.

Parameters from the Architect's Rendering

Windows provide an example of what can be discovered from the architect's rendering. The style of the windows and their relative positions can be seen. They could be single or double hung, casement, bay, or picture windows. If they are single hung, double hung, or casement installation should be simple and rather commonplace. However, suppose one is a bay window, then many questions need to be answered. Are there to be special framing requirements? Probably. The framing is not usual because of the many angles involved with bay windows. There may also be some roof framing as well. The angles will be difficult for an apprentice to cope with. Some journeymen also have considerable problems framing the roof over a bay window. So you can see, windows and their shapes give a house some of its character and immediately bring forth mental images of the structural needs for proper anchorage and inclusion.

What else may be gleaned from an architect's rendering? Doors often play a significant role in a house's character. The door or doors may be any of a variety of designs, usually solid and quite heavy. More significant, though, is how they are framed, and this leads directly to thinking about how they will be supported. Perhaps a colonial style door assembly is used as in Figure 1-1. The wide moldings around the outside of the frame require special care and handling. The trim also poses problems involving flashing and siding, and some thought must be given to the best method for installing the unit so that it is plumb and level, well anchored, securely braced, and waterproofed for many years.

Other features of the house that can be obtained from the rendering are the type of siding and cornice, the porch and its features, the carport, and the breezeway, to name a few. As each part of the house is thought of, the person viewing the rendering should consider the normal structural requirements, as well as the usual tasks, and, if possible, the sequence of construction. If a part of the house is observed but the underlying construction can be visualized only piecemeal if at all, a need for learning exists.

Figure 1-1 Colonial door assembly.

Parameters from the Blueprints and Specifications

To continue, suppose the house is to be framed and have a brick veneer. How would the frieze boards be installed? Figure 1-2 illustrates a detail drawing from a blueprint that gives this data.

Figure 1-2 Details of blueprints reveal much of the final look plus how the house is fabricated.

Figure 1-2 *(continued)*

This leads to the second method of developing the character of the exterior of the building; the use of blueprints and specifications.

Much can be learned from a detailed study of these plans. They can be used to develop a variety of images of the character within the house. For example, the shape of the perimeter is readily seen from the floor plans. This should convey an image of breaks in wall lines, interior and exterior corners, window placement, type of wall to be constructed, parts to be attached, such as porches, breezeways, and carports, and entranceway details.

From the elevation drawings each of the house's sides can be seen, studied, understood in terms of how common, selected, or special principles of construction are applied. In addition, detail drawings can also provide a vast amount of data that relate to the character of the house. From these drawings and the elevation drawings, a mental picture of the roof line can be developed. The roof design can be determined to be a simple *gable* or *A-framed*, *hip* or *hip and valley*, *flat roof*, *gambrel*, or *Mansard* type as Figure 1-3 illustrates. Also, determine if the roof has dormers. If it does are they gable or shed style? These designs mean a great deal in terms of the principles of construction. Chapter 2 covers the designs of roofs in more detail, but for now the idea of developing a mental image of the construction involved should clearly point to using trusses or building the roof with individual rafters and braces. A look at the plan will quickly define the type of construction.

From the outside view, a mental image should be developed of where, for example, hip and valley rafters are needed, where the jack rafters are to be installed, where common rafters should be used, and finally where cripple rafters might be needed. The roof design and shape will tell all of these things. The roof design should also show the type of cornice used. This feature influences the layout of the rafter. If the cornice overhangs the house, the rafter length must be increased to allow for construction of the cornice. If the cornice is the *close* variety, as shown in Figure 1-4, then no rafter overhang is needed. However, proper backing for the cornice must be used and this data is seldom given in plans or specifications.

Where the roof is attached to the wall as in a breezeway, the intersection of roof and wall requires precision installation of roofing shingles, flashing, and wall siding to ensure a sound, waterproof job. After visualizing this picture one should develop mental images of the construction sequence. For that matter, thinking could go into more detail on such matters as the method of tying the rafters to wall framing, sound application of sheathing, and preparation of the flashing materials.

Considerable information on the character of a house can be obtained from a close study of the exterior views. Some of the most prominent features such as windows, doors, siding, roof, and cornice allow the carpenter's mind to delve into the structural

Figure 1-3 Determine how rafters and bracing are included by mentally constructing the roof.

needs that make these features possible. Obviously not all parameters have been developed in these few pages. More information
can be obtained by study of the plans, but there is another avenue
open to the carpenter that can develop a mental image of the
structure.

Roof sheathing

Crown molding

Frieze

Rafter

Wall sheathing

Stud

Figure 1-4 A mental picture of the factors that will affect cornice installation should be made.

Parameters from the Styles of Houses

Although the definitions of the specific character of a house
are fundamental to the builder, there is some regularity of construction if the style of a house can be identified by its overall
appearance. Figure 1-5 shows four different styles of houses. Each
can be built with standard building materials; yet certain characteristics are more predominant in one than in others. For example, the Cape Cod style (Figure 1-5a) seldom if ever has much
overhanging cornice. This is understandable, because in the Cape
Cod region of the northeastern United States the sun helps heat
the house during the long cold winters. In contrast, the ranch
house (Figure 1-5b) has a large overhang so that the sun is blocked
from the walls and windows during the long hot summers of the
south and southwest.

The split-level style of house (Figure 1-5c) came into being
after World War II and found wide acceptance on sloping property. It may be styled for the North with small overhangs or the
South with large overhangs. The two-story house (Figure 1-5d) has
always been an American home and is the most economical house

to build and maintain on a per-square-foot basis. Heat from the lower floor drifts upstairs, roof lines are usually simple and dormers are frequently included.

Figure 1-5 Styles of houses tell a story of their construction. *(Photos Courtesy of Boise Cascade Corporation)*

Figure 1-5   *(continued)* Styles of houses tell a story of their construction. *(Photos Courtesy of Boise Cascade Corporation)*

With these particularly distinctive characteristics, the carpenter can create mental images of the structural needs for floors, walls, window and door installation, as well as rafter and cornice needs. These are important objectives that develop much understanding of the principles used in constructing the exterior of a house.

DEFINE INTERIOR PROPORTIONS AND SPECIFIC CHARACTER

Several concepts can be developed about the interior character of a house from the plans and specifications. Room sizes and arrangement are the most obvious. Why, though, are these rooms placed and sized as they are? The lifestyle of the family, number of members, and family income probably dictate much of the size and location of the rooms. However, these values frequently are secondary to the architect that designs the house. He is concerned with other, and to him more important, considerations. He knows that more people can afford homes if they are built economically, with a minimum of waste materials, and yet meet FHA and other national building codes. He therefore selects a group of rectangularly proportioned rooms and arranges them into a house. Using this method he maximizes use of all materials, provides a general purpose house to meet the general needs of the average family. Once the basics are established, he applies trim and other individual features much as auto manufacturers do when dressing up cars. Some rooms may be paneled, others papered, and some with chair rails for individuality and sales appeal.

Parameters for Bearing Walls and Proper Corner Construction

The carpenter, on the other hand, must view the drawing and specifications for other purposes. Namely, are there load-bearing or non-load-bearing partitions? If they are load bearing it means that ceiling joists will rest on top of them. It also means that if the bearing partition runs parallel to the floor joists, trimmer joists or double joists need to be installed in the floor. It also could mean that a girder needs to be installed beneath or within the floor joist assembly. A non-load-bearing partition does not need reinforcement to the extent a load-bearing partition does. Some reinforcement such as solid bridging should be included in both floor and ceiling joist assemblies.

Another aspect of interior character of a house is that all wall materials meet at corners and along the ceiling. So the need for proper backing to nail or bond the wall materials must be provided. A study of the plans should reveal the partitions that run perpendicular (at right angles) to the ceiling joists need no additional ceiling-corner backing, but those that run parallel to the ceiling joists do. Interior side-wall corners need backing for both types of walls, and assemblies similar to those shown in Figure 1-6 must be built and installed to provide the adequate, sound backing needed.

Figure 1-6 Sound anchorage for walls and wall coverings is not found in plans, so it must be thought of by the carpenter.

Parameters for Sound Control

Television, radio, stereo, and a need for quiet are all important parts of the average family's daily living. To provide these for the individual, and yet allow others in the same household privacy for their needs, rooms are increasingly being soundproofed. The specifications could call for sound control at STC 50, for example. This level of sound control means that regular levels of noise will not penetrate the wall or radiate from room to room. However, sharp sounds, such as a chair falling over or a book falling to the floor, may be audible. Soundproofing is accomplished in many ways; two that can be seen from plans are the *double-stud double plate* (Figure 1-7a) and *double-stud single plate* (Figure 1-7b) construction techniques. Other data taken from specifications can call attention to other techniques. As Figures 1-7c and d show, resilient channels, sound deadening board, wool or glass fiber insulating batts and carpeting may be required

Figure 1-7 Understanding the principles of sound control through alternate construction principles and methods.

to achieve desired results.[1] Other sources of information on sound control must be studied to learn what is needed for a thorough and effective job.

Parameters for Modular Wall Panels

Finally, the character of the interior may be set by modular wall panels. These are wall panels varying in thickness from 1-1/2 inches to several inches thick and are finished on both surfaces. They are usually set into channels that are fastened to the floor and ceiling. There installation and removal is simple and quick. These walls could be installed in residential houses where truss roofing is used. Since the truss is self-supporting, the lower chord (same place as the ceiling joist) acts as a ceiling joist, and partitions may be installed anywhere and at any time. All the interior partitions are non-load-bearing.

Several constraints are placed on moving walls with wiring and plumbing. A study of the plans and specifications should reveal if modular walls are included, and if they are a procedural sequence of construction can be developed. For example, blocking between trusses can be installed where partitions will initially be set; then the entire ceiling of the house may be covered with wall board. All perimeter walls should be wired and have the plumbing installed, then insulated and wall-boarded. Finally, modular wall units as partitions can be installed.

There may be other methods of developing the character of the house's interior, but all should ultimately lead the carpenter to the application of construction principles to meet any and all situations.

SUPPORT REQUIREMENTS

Thus far this chapter has illustrated the need for the carpenter to study closely the pictorials, plans, and specifications of a house to develop mental images of each completed segment. To this end, examples were given to help you understand how to relate what you see to what is needed during construction. There are several

---

[1] For more on sound control see Chapter 6 of *Carpentry in Commercial Construction*, (Reston Publishing Co., Inc., Reston, Va.).

additional and fundamental requirements that must also be made a part of the image. These are the support requirements for floor and ceiling openings, the alignment of joists, studs, and rafters and the slope of the roof.

Openings in Joist Assemblies

A structural deficiency exists wherever an opening is made in a floor or ceiling joist assembly, whether it be for a chimney, stair, or fan. Additional materials and strengthening methods must be used to overcome the deficiency. As a rule, headers and trimmer joists as shown in Figure 1-8 are employed to reinforce the structure. The headers support the free end of the joists, the double header provides needed strength. The trimmer joist provides support in a parallel direction to the joist and provides sound anchoring for partitions or other construction forms that may be attached to the opening.

Figure 1-8 Understanding the principles necessary to overcome structurally weak areas.

(b)

Figure 1-8 *(continued)* Understanding the principles necessary to overcome structurally weak areas.

Pressure and Force from the Roof

Figure 1-9 shows the direction and points of force that are present in any house. Notice that the rafter force is directed to the plate. Through the plate and into the stud is a downward force that continues through the floor joist to the foundation. It is therefore vital that the rafter and wall stud be aligned one on top of the other and that the studs align over the floor joists. Should the rafter be offset from the stud, its force and weight may be sufficient to shear the plate members. Some types of lumber have extremely low shear values.

The support requirements of the roof, that is, the ability of a roof to carry its normal dead load of plywood and shingles as well as a snow load, depends to a large extent on its slope. The greater the slope the more the force is directed to the rafter-wall-plate junction and therefore less direct weight is applied to the rafters.

However, in a low-sloping roof, say a 4-in. or 5-in. pitch, the weight of snow would cause a more direct downward force, and thus possibly cause a structural collapse. So what must be done?

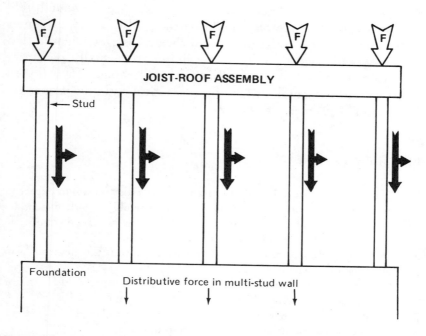

Figure 1-9 Forces always exist, but how to cope with those developed by the roof?

The members selected for rafters must meet all of the requirements to support both the dead load, a live load, and a snow load, where applicable. To this end, certain species of fir or pine are used. They must be properly cured and seasoned and must be of appropriate size. Tables giving these parameters are provided in the appendix.

From this brief discussion, it is clear that a study of the related factors including physical layout, geographical location, snowfall predictions, slope of roof, and materials of sufficient quality and strength is imperative to the carpenter to ensure that the house is soundly constructed.

In conclusion, the ideas put forth in this introductory chapter were developed from the experience of many master carpenters. The best of the trade can and do mentally develop a sequence of construction long before the first board is selected. The best men know the exact sizes of each piece of lumber, where it goes and when to install it. They know where the points of force are the greatest and what steps are necessary to ensure a sound, reliable structure. It therefore should be the goal of every carpenter to become a master craftsman, and if the goal is ever achieved that carpenter will have developed the ability to mentally construct each segment of a house as well as the whole house. He will then translate those mental images into practical tasks in proper sequence, and thus create a beautiful work.

PROBLEMS AND QUESTIONS

1. By studying Figure 1-1, state the framing materials you would expect to use to frame the opening. Assume the door assembly is 5 ft. 6 in. wide.
2. Study Figure 1-2, the floor plan, and determine the number of 2 x 4s you would use for outside corners on the exterior walls.
3. What problems would you expect to have to solve if the elevation plan called for a bay window?
4. Referring to Figure 1-2, the elevation drawings, what is the exterior covering of the building? What is the pitch of the roof?
5. Why does the frieze board in Figure 1-4 have to be rabbeted?
6. Describe two methods of framing a soundproof wall.
7. Which style of house uses:
   a. Close cornices   b. Large overhangs
8. Define a load-bearing wall or partition.

2

Designs of
Residential Houses

Chapter 1 was devoted to explaining how the experienced carpenter or master craftsman visualizes all of the different tasks he will perform. Subjects were generalized, but examples of portions of homes were selected to show how and why trained carpenters become so. In this chapter the approach is more selective and focuses on discrete designs of homes found throughout the country. Each of the subjects represents a particular style or design of home. Most frequently the design resulted from environmental conditions, social attitudes, and availability of resources.

In the development of the colonial home design, most included a porch or veranda that was used socially during the cool mornings and evenings. The porch ceiling and its roof also provided shade for the side walls of the home, thereby acting as an insulator between outdoors and indoors. Large windows and high ceilings also contributed to comfortable living. Consider also that the houses were not insulated and had no central heat or air-conditioning.

Contrast the other designs with the colonial and they too have sound reasons for their design. The two story without colonial styling is beautiful, not at all commonplace, but is a very excellent design. The ranch style is a rambling, spread out design and its uniquely large overhanging cornices provide protection from the hot summer sun. The split level is extremely popular in hilly country because of the added rooms that may be obtained without altering the terrain or resorting to extensive land fill or excavation. Then too, the saltbox and Cape Cod styles are simple in design. They were originally designed to make use of the sun during the year round. Finally, the contemporary or ultra-modern style evolved along the coast line and in special mountain areas where views are exceptionally beautiful. Its design maximizes the use of natural light and attempts to integrate the indoors with nature.

This brief introduction to housing designs narrows the selections that follow and focuses attention not only on style but also on the simplicity or complexity of design that translates to specific tasks that must be done when constructing any one of the houses.

CONSTRUCTION FEATURES OF HOUSING STYLES

The Southern Colonial Style

One of the main characteristics of the southern colonial design is the decorative columns that support a porch or pediment. Frequently these are round hollow assemblies with bases and crowns, but they may also be square. Figure 2-1 shows four round columns. When columns like these are used they are purchased from a wood mill and the carpenter merely installs them. Square columns, on the other hand, are quite frequently made on site.

Figure 2-1 The southern colonial style. (*Courtesy of Boise Cascade Corp.*)

Figure 2-1 *(continued)*

The roof design in Figure 2-1 is the simplest of all designs, the gable. The main roof over the two stories and porch or pediment extends beyond the columns and all sides to create wide overhangs. Since the roof line is unbroken it may be assumed that ceiling joists extend out over the porch and are used as part of the roof assembly and porch ceiling. On the other hand, simple truss assemblies may be used quite easily for the entire structure.

Trusses may also be used for the lower gable roofs or they may be made from properly designed and braced rafters. The lower roofs also carry the wide overhang design.

All cornices are *closed* or *boxed* in. The *rakes* and *soffits* are all enclosed with plywood. Although not shown, ventilating panels or screens are installed in the soffits to allow proper air flow in the attic.

The house is wood framed with the probability that *platform framing* is used rather than *balloon framing.* The reason is the unbalanced window and door layout of the front wall. Notice that there are different window widths on the second floor when compared with the first. When balloon framing is used, windows on each floor are usually the same width and are aligned vertically. Another reason for using the platform framing technique is that the upper floor extends slightly out over the lower wall.

Lap siding is used to cover the outer walls except where brick veneer is installed. This creates a special problem for the carpenter. He must make sure that the siding brick joints are well sealed with caulking and overlap in the direction of rain fall.

The Dutch Colonial Style

The Dutch colonial style employs the gambrel roof design seen in both homes and barns. The home shown in Figure 2-2 makes use of the small section of gambrel roof to add the style. This roof is very complex in design. The main roof is a gable frame design. Where the gambrel section ties in, valley rafters and valley jacks are needed for framing the upper level of the gambrel and main roof. The lower, steeper sections of the gambrel are framed and attached to extended floor assemblies and side wall. Overhangs are large but straightforward in design.

Figure 2-2   The Dutch colonial style. (*Courtesy of Boise Cascade Corp.*)

Framing the front wall is somewhat complicated because of the extension under the gambrel roof (right) and the projected upstairs wall that contains the bay window (left). In addition, the entrance door assembly is recessed two feet from the outer wall line.

Siding is of the lap variety, but rather than using self-lap on corners, vertical corner boards are used. This makes all joints butt and also means that tar paper or vinyl should be installed behind these corners and joints to prevent leaks.

Aside from the special considerations listed above, the house is easily built using the western or platform framing technique.

The Simpler Colonial Style

Not all colonial designs are elaborate or ornate as Figure 2-3 shows. Here is one of simpler lines, yet still retaining a colonial character. The porch across the front protects the entrance and provides a place to socialize. The four columns support the porch roof and its ceiling. The porch ceiling joists extend from ceiling joists of the house.

Figure 2-3 Simple colonial styling. (*Courtesy of Boise Cascade Corp.*)

The roof is a simple gable design with ridges at two different levels. The center section of roof uses longer rafters to allow for coverage of the porch roof; yet the rafters along the entire rear of the house are uniform in size and are in alignment. Overhangs are generous and finished with *soffits* supported by *fly rafters, ledgers,* and *lookouts.*

The frame of the house is rectangular with no breaks in any walls. This simplifies construction. Yet, because of the roof line it appears as if there are breaks in the line of the front wall. Siding is lap with self-corners. Windows are used in unbalanced design.

The Country Classic

The country classic design shown in Figure 2-4 is very complex. It has many interior and exterior corners and its roof is especially complex. A portion of the house is two story and the remainder is one story.

Figure 2-4 The country classic style. (*Courtesy of Boise Cascade Corp.*)

The roof, for example, employs a gable with valley design. The main roof over the second floor employs two sideless dormers. The lower roof extends along the porch and develops a valley with the roof over the short extension to the right. This arrangement complicates construction. The garage roof is a gable that joins the main roof. It is probable that equal pitches are used for both roofs; but, because the spans are different, the ridges do not align. Or it could be that an unequal valley pitch is created. Elevation plans would need to be used to determine the pitches of the different roofs. The fireplace in the rear wall will require special carpentry work in the roof as well as wall and floor framing.

The columns may be built on site; if so, they should challenge the craftsman carpenter. Railings with spindles complete the porch.

Siding is lap with corner boards used throughout. Again, special attention must be paid to corners and siding installed this way.

The Two Story Using Balanced Design

The two story shown in Figure 2-5 is one of balanced design insofar as the windows on each floor are identical to those on each side of the entrance and the entrance is a balanced design.

Figure 2-5 A two-story house with balanced design. (*Courtesy of Boise Cascade Corp.*)

Figure 2-5 *(continued)* A two-story house with balanced design. *(Courtesy of Boise Cascade Corp.)*

Its roof is a simple unbroken line of gable design and may easily employ trusses. Western or platform framing should be used because of the cantilever projection of the upstairs rooms.

Brick veneer on the lower walls requires the carpenter to extend the siding over the brick. Again siding is lap with corner boards used.

A simple framing technique prepares the opening for the front door assembly. Notice that it is recessed slightly and that the recessed line carries through to the roof. As on several of the other homes staircases need to be constructed inside.

The Two Story Using Balanced Design

The two-story design shown in Figure 2-6 truly has a balanced design on the main section. The windows are the same size and are equidistant either side of the entrance. The roof is a simply designed gable with finished overhangs. There is a fireplace at the rear, so the roof will need framing around it as will the walls on either side.

The porch is also a simple gable but its ceiling is not so simple to construct. It is elliptical. This means special ceiling joists need to be cut. The ceiling materials will then be anchored to these specially prepared ceiling joists. In all probability 1/4-in. plywood or hardboard will be used. The porch roof is supported by four columns that may be fabricated on site. Lapsiding is butted to corner boards. Gable ends are finished in plywood, which breaks up the line created by the lap siding.

Inside the house a staircase is built that requires a special stairwell framing in the second-floor frame. The staircase is not too difficult to construct since it has no turns or curves.

Figure 2-6 Another two-story house with balanced design. (*Courtesy of Boise Cascade Corp.*)

The Split Level

Designed for sloping land, the split level shown in Figure 2-7 is constructed with the platform framing technique. It has a simple gable roof with wide finished overhangs.

The lower levels are brick veneered and this means that the carpenters must fashion the *frieze boards* to overlap the top course of brick. This also means that the upper story must project out over the lower walls to accommodate the brick veneer.

Siding used on the upper level reduces the weight on the footings and also provides a contrast in the lines of the design. A partial balanced design is used on the house where the windows of the two stories are of the same sizes and are in vertical and horizontal alignment.

Two short sections of stairs are needed to allow passage from each level.

Figure 2-7   A split-level style. (*Courtesy of Boise Cascade Corp.*)

The Split Level with Spanish Accent

There are several differences between the split level shown in Figures 2-7 and 2-8. They are the porch that requires a roof and the vertical siding on the upper level; both are shown in Figure 2-8. Aside from these changes the houses are alike. Therefore, the comments about structure made before apply to this house except for the additional roofing required to cover the porch and vertical siding and plywood gable end covering.

The rafters are extended to cover the porch, but notice that the line remains unbroken; that is, all rafters are in alignment and a single ridge level is used. There are two building requirements to note though. First, the framing above the garage door is higher than normally expected because of the shortened rafters. Secondly, there is a small valley in the porch roof over the entrance side of the porch.

This valley sheds water from the upper roof away from the wall of the upper level thereby reducing the risk of leaks. It also causes the water to flow from the lower roof to the shed and into the garden, not on the entrance sidewalk.

Figure 2-8 A split-level style with Spanish accent. (*Courtesy of Boise Cascade Corp.*)

Figure 2-8 *(continued)* A split-level style with Spanish accent. *(Courtesy of Boise Cascade Corp.)*

The vertical siding is board and batten. Notice that it does not extend to the rakes of the roof nor does it extend to the bottom of the wall. Based on these observations one condition exists for flashing. It is on top and behind the lower ends of the lower board band and vertical siding. The vertical plywood panels used for gable ends must extend over the board and batten siding if the possibility of leaks is to be eliminated. If the plywood were nailed directly to studs, then flashing at the top of the board and batten siding would also be needed. Finally, notice that beams are visible extending under the upper level. These may be just for show and are attached rather than part of the support structure.

The Ranch Style with Carport

The ranch style house shown in Figure 2-9 is a one-story home constructed around a basic rectangular plan. There are no breaks or offsets in the wall designs so construction is very straightforward.

The roof is a basic gable type with large overhang that is finished in a *boxed* or *closed cornice* style. A girder will be required across the carport opening to support the roof. Truss roof assemblies could easily be employed for the roof's construction. The dormer is of the sideless variety and adds a dimension of style.

Vertical board-and-batten siding covers the exterior walls except on the gable ends under the main roof. Here exterior plywood panels are used.

Figure 2-9 A Ranch style house. (*Courtesy of Boise Cascade Corp.*)

The columns on the carport that support the roof are probably square and made on site. They may be 4 x 4s or doubled 2 x 4s cased with 1-in. stock, trimmed at the base and crown.

The Ranch Style with Single-Car Garage

The basic structure of the ranch style house shown in Figure 2-10 is rectangular on a single level. No offsets or breaks in the walls are employed, which simplifies construction. The garage is framed along with the house and a header is used across the garage door opening.

Figure 2-10 A modest ranch style with garage. (*Courtesy of Boise Cascade Corp.*)

The roof design is a basic gable style with a few of the rafters extended to cover the porch. There is a fairly large overhang in *closed cornice style* that adds a bit of formality. The porch roof is supported by three simply designed posts. The severity of straight lines and plainness is broken by the inclusion of curved frieze pieces between posts. Notice that the porch ceiling slants along the roof line. This is necessary to provide head room for the windows and the door.

Brick veneer is used as the exterior wall covering for the front of the house. This means that the frieze board under the cornice soffit must be fabricated away from the wall about 4-1/2 to 5 in.

Vertical siding covers the remainder of the house including the gable ends. Board and battens are used, and on the gable ends the battens must line up with those on the wall if the line is to look neat and proper.

The Two-Story Saltbox

The two-story saltbox style house shown in Figure 2-11 is of balanced design. Windows are aligned over each other and are of the same style. The door is centered in the front wall. A breezeway connects the two-car garage to the house.

Figure 2-11 A two-story saltbox style house. (*Courtesy of Boise Cascade Corp.*)

The roof is somewhat complex because of the breezeway and elongated rafters on one half of both the main house roof and garage roof. This is typical of this style house, however. In many earlier homes of this style, rooms were added as the family increased in size. The roof line was retained to cover the new additions at the rear of the house.

The main roof is a gable type with overhangs to the front and back but uses *closed* cornice design on the gable ends. Here facia boards cover the lap siding. The breezeway roof ties into the side of the house and garage. Where it ties into the house flashing must be used for leak-proof installation. Where it ties into the garage roof valley rafters and valley jack rafters are used.

The garage roof uses rafters of two different lengths. Those on the breezeway side are shorter than those opposite. The amount of overhang at rafter ends equals those of the main house as do the *closed* cornices on gable ends.

A cantilever projection of second-floor joists is made along the front wall where a projection of several feet is shown. The extended ends of these joists must carry the load of the wall and roof so special care needs to be taken during construction. Sound design as well as good quality strong lumber needs to be used.

Lapsiding is used on all the walls with corner boards installed first.

The breezeway uses four simply designed columns and they are incorporated in the design of the porch. The frieze board along the ceiling level of the porch is fashioned in a manner similar to the tops of the garage door openings.

The Contemporary

The contemporary style house shown in Figure 2-12 is very complex in design. Many breaks in exterior wall line result in interior and exterior corners. Many different roof segments are used, and some walls are single story while others are two story.

The pitches of the roofs are quite steep; however, this does not pose a construction problem. The fact that they are at different levels and lengths does. Rafters will be custom cut to a variety of lengths. The various lengths do not alter the pitch nor will they change the ridge, bird's mouth, or overhang rafter tail cuts.

Figure 2-12 A contemporary style house. (*Courtesy of Boise Cascade Corp.*)

Much of the house is two story, yet because of the roof design exterior walls will extend above ceiling levels and cathedral ceilings are used in some rooms.

The siding is installed in two styles, diagonally across the front and horizontally on the remainder of the house. Corner boards are used to simplify corner joints where the two directions of siding would be difficult to join. Cornices are *close* on sidewalls and boxed on overhangs in front and rear.

Framing for the clerestory windows between the front roof and higher rear roof needs to be sound to avoid structural problems and to provide a solid base for windows. Windows installed in this area need to be of very high quality and well insulated so that no possibility of leaks is present.

The inside staircase includes a landing. This means that two sections of stairs are to be built and installed. Another short section of staircase needs to be built onto the rear deck. Both sets of stairs are different in design and structure. The interior one is more formal with risers and treads fit into side stringers, all made of fine softwoods or hardwoods. In contrast, the deck stairs are made with step treads and stringers under the treads. No riser boards will probably be used. Further, these stairs will probably be made from treated pine or fir.

LEARNING MORE ABOUT PRINCIPLES OF CONSTRUCTION

The pictorials of houses can provide much valuable information to the carpenter. Sometimes this information is readily translated into tasks that are already understood and other times the information poses problems of two kinds. One, what is the task or tasks that must be performed, and two, what is the way or ways to perform the task?

If the task is readily recognizable, its completion during the construction phase should pose no problem. Further, if a thorough understanding of the principles of construction and design are known, then a full grasp of it is appreciated. Should any slight variations from the commonplace be seen, they too are understood.

On the other hand, if the pictorial does not communicate a task or series of tasks that must be performed, then the carpenter lacks knowledge of the principles of construction and design. A clear need for learning exists. Studies can be conducted in formal classrooms and at home with books.

Most of the designs of houses identified in this chapter are simple in principle and construction. A large percentage of them are rectangular and employ the simple gable roof. Many are built on concrete slabs and are single story. Therefore, the western or platform framing technique is used. Modular construction is also possible. Balloon framing is seldom used. Even though cornices are boxed they are plain and few if any problems are expected during their construction.

Several of the houses do, however, employ complex roof and wall design. These features relate to design and construction principles that require considerable study. Understanding the hip-and-valley roof design requires a knowledge of force, loads, principles of design, and adaptation of the tools to be used. Several chapters to follow devote considerable attention to roof design and planning. Walls must also be constructed soundly so that forces from the roof, joists, winds, earthquakes, and so on are predictable and accounted for. For instance, how does a header carry its load and distribute it to the studs? Or, what size header must be used to carry a load of a certain size? Questions like these are a part of learning and understanding the principles of construction and design.

Another example where adaptations of tasks must be used to employ a single principle is installing flashing. The principle requires a metal strip be formed with a bend and installed so that the natural fall of rain sheds down its surfaces. If, for example, the tops of windows or door frames need flashing, these pieces extend up under the siding and out and over the window or door top. On the other hand, if flashing is used where a roof intersects with a side wall as shown in Figures 2-1, 2-2, 2-3, and others, small pieces of flashing are installed on top of each row of shingles. Afterwards the siding is installed on top of the pieces of flashing.

There are literally hundreds of different tasks that have many subtle or large variations on both simple and complex principles. To name or list them would be useless. They cannot be found in a set of plans or specifications either. They must be learned on the job and from study. Many of the chapters that follow are developed for the single purpose of providing a clear and complete understanding of the principles employed in various tasks so that the carpenter knows what he is doing is correct.

PROBLEMS AND QUESTIONS

1. What is a distinctive feature of a southern colonial house?
2. Look at Figure 2-1 and determine the type of roof each section uses. What type of rafter is used?
   a. common   b. hip   c. cripple   d. valley

3. Why would western framing probably be used for the house shown in Figure 2-1?

4. Determine the different types of rafters used in the house shown in Figure 2-2.

5. Does the house shown in Figure 2-2 use self-corners on lap siding or corner boards?

6. What is the purpose of the four columns on the house in Figure 2-3?

7. Cite at least four details of the country classic house (Figure 2-4) that complicate its construction.

8. Does the two-story house (Figure 2-5) have a balanced or unbalanced design? The two story in Figure 2-6?

9. Why would platform framing be used for the split-level house (Figure 2-8)?

10. What would be the difficulty in building the contemporary house (Figure 2-12). List at least three factors.

11. Take a pad and pencil and observe a neighbor's house. Write down as much design detail and construction detail as you can. Have your instructor or a fellow carpenter verify your data.

3

Lumber — A Key Building Material

In its various forms, lumber is no stranger to the carpenter. It has almost limitless uses in building construction. To name a few, it is used in making forms for concrete, framing, trim, and wall coverings.

Because this material is so versatile and common, many studies have been conducted to determine its characteristics. The results of these experiments are beneficial to every builder. This chapter defines and examines several of the most important characteristics of lumber to the builder. First we identify the two types of woods and some species of each type. Grading names and definitions are presented next. Wood properties, which play an important role in sound construction, are then defined and several examples of their use are provided.

TYPES OF WOOD

Each species of wood is placed in either the *softwood* or *hardwood* group depending on the type of tree. By definition, the softwood tree is cone bearing and has needles or scalelike leaves. In contrast, the hardwood trees usually drop their leaves in autumn.

Softwoods

Many properties of softwoods are shown in Figure 3-1. Notice that the area of *early wood* (5) is much larger than late wood (6). Also notice where the resin duct is located. With some imagination you can visualize the grain of a board as well as the

Softwood key
1. Cross sectional face
2. Radial face
3. Tangential face
4. Annual ring
5. Earlywood
6. Latewood

7. Wood ray
8. Fusiform ray
9. Vertical resin duct
10. Horizontal resin
11. Bordered pit
12. Simple pit

Figure 3-1 Wood structure of a softwood. (*Source:* USDA, PA 900.)

end cut. The most common softwood species used in house construction are the firs, pines, cedars, hemlocks, larches, and spruces. Table 3-1 lists many of the different kinds of each species:

TABLE 3-1   Species of Softwoods

Fir	Pine	Larch
Douglas fir	White	Western
Silver (pacific)	Yellow	Alpine
White	Ponderosa	American or
Noble	Sugar	black
Red (Calif.)	Loblolly	Tamarack or
	Short leaf	Hackmatack
	Slash	

Hemlock	Cedars	Spruces
Western	Port Arford	Sitka
Eastern	Alaska	Engelman
	Incense	White
	Western red	Black

Many of the species listed are used for framing. Some, such as the douglas fir and southern pine, are very good. Others that are used but do not have the strength of these two are the hemlocks, spruces, cedars, and some pines. More discussion of the individual qualities of each species occurs later in the chapter.

Hardwoods

Figure 3-2 shows several properties of hardwoods. For example, the sizes of early and late wood cells are both small and almost equal in area. This characteristic accounts for the density of hardwood. The closeness in annual rings also accounts for the fine grains usually found in hardwoods and explains why fine end cuts can be made.

Hardwood key
1. Cross sectional face
2. Radial face
3. Tangential face
4. Annual ring
5. Earlywood

6. Latewood
7. Wood ray
8. Vessel
9. Perforation plate

Figure 3-2 Wood structure of a hardwood (*Source:* USDA, PA 900.)

The number of hardwood species used in construction is limited to birch, elm, maple, oak, mahogany, and walnut. These woods are seldom used for framing. However, they are used for paneling, cabinets, doors, trim, and molding wherever stained woods are desired for their beautiful grain and color.

GRADING OF LUMBER

Every builder needs to know and understand how lumber is graded and what the grades mean. The grading rules are established and modified by the lumber-producing companies, and associations of these companies, in coordination with the U.S. Departments of Commerce and Agriculture.

Through the efforts of committees from industry and government a fair set of standards has evolved. Naturally, the grading must be equitable so that the user can rely on minimum quality and structural capabilities of the lumber, yet provide a source of income and profit for producers.

The Various Grades

There are grades established for all types of lumber. Some of the more common that are important to the house builder are the *select*, *finish*, and *construction* grades. Table 3-2 provides a brief identification of some of the most commonly used grades.

TABLE 3-2. Grades of Lumber

A. Select Grades of Lumber[a]

No. 1 and 2 clear B/and BTR C select D select	Relatively clear lumber used where finishes allow some imperfections. Uses: trims and cornices.

B. Finish Grades of Lumber

Superior finish Prime finish E finish Casing and base Flooring Clear paneling	The highest quality of finish lumber, often clear and excellent for staining.

[a]Grade titles and listings are those generally used by Western Woods Products Association and many other inspection bureaus.

Table 3-2 *Continued*

C. Common Grades of Lumber

No. 1 common	Available in most sizes, but not al-
No. 2 common	ways in all grades. Frequently used
No. 3 common (standard)	in house construction and often fin-
No. 4 common (utility)	ished with transparent coatings.
No. 5 common (industrial)	

D. Framing Grades—Structural

Dense select structural	These are the grades widely recom-
Select structural	mended for general construction. The
Dense No. 1	list identifies in descending order the
No. 1 and appearance	best to poorest grade of lumber by
Dense No. 2	quality.
No. 2	
No. 3	
Stud	
Construction	
Standard	
Utility	

The Standard of a Grade

Associated with each grade is a standard that sets the types of defects that are acceptable and the limitations of these defects. The inspector-grader first selects the best-looking surface and then applies the rules for grading. Since human judgment is applied, the highest grade is selected rather than the lowest. This is a rather difficult task for an inexperienced person. Consider the need to grade a 2 × 6 joist as *select structural.*

Dimensional lumber of this quality is limited in characteristics that affect strength and stiffness values to provide a fiber stress in bending value of 65% of that allowed for clear, straight-grain wood and to provide a recommended design value for modulus of elasticity of 100 percent of that allowed for the clear wood average. This grade is recommended for use in application where both high strength and stiffness values and good appearance may be required.

Characteristics permitted and limiting provisions shall be:

Checks—Surface seasoning checks, not limited. Through checks at ends are limited as splits.

Grain—Medium.

Knots—Sound, firm, encased, and pith knots, if tight and well spaced, are permitted in sizes not to exceed the following, or equivalent displacement:

Nom. Width	At Edge Wide Face	Centerline Wide Face	Unsound or Large Knots & Holes (Any Cause)	
6"	1 1/4"	1 7/8"	1"	
8"	1 1/2"	2 1/4"	1 1/4"	One hole or equivalent smaller per 4 lin. ft.
10"	1 7/8"	2 5/8"	1 1/4"	
12"	2 1/4"	3"	1 1/4"	
14"	2 3/8"	3 1/4"	1 1/4"	

Manufacture—Standard "E".

Pitch and pitch streaks—Not limited.

Pockets—pitch or bark—Not limited.

Shake—On ends, limited to 1/2 the thickness. Away from ends, several heart shakes up to 2 ft. long, none through.

Skips—Hit and miss skips in 10% of the pieces.

Slope of grain—1 in 12.

Splits—Equal in length to the width of the piece.

Stain—Stained sapwood. Firm heart stain or firm red heart limited to 10% of the piece.

Wane—1/4 the thickness, 1/4 the width. Five percent of the pieces may have wane up to 1/2 the thickness and 1/3 the width for 1/4 the length.

Warp—1/2 of medium.[1]

Clearly, all the different characteristics of trees are included in establishing a standard for a single grade of lumber. Each of the

---

[1] *Source:* Grading Rule for Western Lumber, 3rd ed., p. 101-103, *courtesy of Western Woods Products Association.*

different grades has a unique set of standards, including most if not all those listed for select structural. Of course, the allowable defects are stated differently.

WOOD PROPERTIES IMPORTANT TO THE BUILDER

Lumber cut from any tree contains many individual and common properties. For example, a 2 X 4 may have been cut from sapwood or heartwood, may be strong or weak, may have internal compression or tension, and may be green or dry. The importance of these properties is developed and is fundamental to future studies.

Sapwood versus Heartwood

Sapwood is made up of the outermost growth rings of the tree. It is usually the lighter colored wood and contains some resins and gums. Heartwood, by contrast, is the dead tissue surrounded by the sapwood. Its only function is to support the tree. The heartwood is frequently referred to as *pith*.

Heartwood is generally more durable than sapwood because it resists decay. On the other hand, it is usually brittle and thus, tends to have higher shear values. Sapwood, although less durable, can be chemically treated more easily. Sapwood also has greater tensile strength because of its superior flexibility.

Specific Gravity

The specific gravity of each species of wood is the measure of the amount of cell wall materials in the wood. It thus is a useful factor for determining the strength of the wood, which then aids in selecting it for various purposes. To cite several examples, consider that douglas fir, southern pine, and oak all have high specific gravities. This indicates that they are all suitable for framing materials where many forces, stresses, and weights must be dealt with. On the other hand, ponderosa pine, spruce, and other lighter woods with lower specific gravities are not desirable where load bearing is expected. They are desirable, however, for trim and other non-load-bearing uses.

The method for determining the specific gravity of a species of wood is as follows:

1. A block of wood $1 \times 1 \times 10$ in. is cut and dried in an oven at 105°C until a constant weight is determined.
2. A volume of water equal in weight to the dried wood specimen is poured into a tall glass cylinder (the shape of the cylinder forces the block of wood to float upright).
3. The dried wood block is lowered into the water to the point where it floats, then removed.
4. The specific gravity is found by measuring the length of wood under water (which is the weight of the wood) by the weight of equal volume of water.

Let us assume that the block sinks 3 in. into the water. This means that the weight of the wood is supported by the force of $1 \times 1 \times 3$ in. $= 3$ in.^{3} of water. *Therefore:*

$$\text{Specific gravity} = \frac{\text{weight of object}}{\text{weight of equal volume of water}}$$

$$= \frac{3 \text{ in.}^3}{10 \text{ in.}^3} = 0.3$$

Table 3-3 lists some of the softwoods and hardwoods and their specific gravities so that a comparison may be made of their relative strengths.

TABLE 3-3   Woods and Specific Gravity

Hardwoods	Spec. gravity	Softwoods	Spec. gravity
White oak	0.57	Western larch	0.51
Elm	0.57	Douglas fir	0.48
Beech	0.56	Short leaf pine	0.46
Birch	0.55	Western hemlock	0.38
Black walnut	0.51	Ponderosa pine	0.37
Bass wood	0.32	Sitka spruce	0.37
		Western white pine	0.36
		White fir	0.35
		Incense cedar	0.35
		Engleman spruce	0.32

Table 3-3 shows considerable range in the specific gravities of both softwoods and hardwoods. It would appear that more hardwoods are suitable for construction of houses. This is true, but fortunately several varieties of softwoods are equally strong and suitable for this purpose. It is also fortunate that softwood trees outnumber hardwood trees by millions to one, thus making lower housing costs possible.

Modulus of Elasticity

By definition, the *modulus of elasticity* symbolized as (E) is the relationship between the amount a piece of lumber deflects and the size of the load causing the deflection. In simpler terms, the modulus of elasticity is the measure of a board's stiffness. Inherent in the definition is the understanding that the member (board) must recover its original shape and size after first being strained and then having the force removed. Although each species of lumber has an assigned modulus of elasticity, the values have been assigned to the various grades of lumber. This simplifies use of the information when selecting joists, rafters, and other structural members.

Table 3-4 provides a distinct relationship of the modulus of elasticity to the grade of lumber. As shown, a 2-in. piece of lumber 4-in. or wider graded as *dense select structural kiln dried* has a modulus of elasticity   of *2 million pounds per square inch*. In contrast a utility stud (2 × 4) only has a modulus of elasticity of 1.4 million psi. Since modulus of elasticity is related to stiffness, the clear interpretation is that a utility stud does not have the quality of stiffness of a stud graded Dense Sel Str KD.

Tables containing modulus of elasticity data by species and grade will be presented in later chapters. In these tables the values of modulus of elasticity vary from those in Table 3-4 and may in some cases be less than 1 million psi. Species with modulus of elasticities less than 1 million psi are not recommended for use in load-bearing structures because it is doubtful that they could recover their original shape after being placed under heavy stress.

Determination of the allowable stress in lumber cannot be made from the E value alone; its F_b or extreme fiber in bending rating must also be considered.

TABLE 3-4: Allowable Stress by Grade of Lumber

| | 2-4" THICK, 6" x WIDER | | | | 2-4" THICK, 2-4" WIDE | |
| | SPIB Grdng. Rule Par. No. | Extreme Fiber In Bending F_b psi / 1.15 F_b psi | | Modulus of Elasticity E psi | Extreme Fiber In Bending F_b psi / 1.15 F_b psi | SPIB Grdng. Rule Par. No. |
Product Grade		F_b psi	1.15 F_b psi	E psi	F_b psi / 1.15 F_b psi	
Dense Sel Str KD	237	2250	2600	2,000,000	2650 3050	210
Dense Sel Str	236	2100	2400	1,900,000	2450 2800	209
Sel Str KD	235	1950	2250	1,900,000	2250 2600	208
Sel Str	234	1800	2050	1,800,000	2100 2400	207
No. 1 Dense KD	241	1900	2200	2,000,000	2250 2600	214
No. 1 dense	240	1800	2050	1,900,000	2050 2350	213
No. 1 KD and Appearance KD	239	1650	1900	1,900,000	1900 2200	212
No. 1	238	1500	1750	1,800,000	1750 2000	211
No. 2 Dense KD	247	1550	1800	1,800,000	1850 2150	220
No. 2 Dense	246	1450	1650	1,700,000	1700 1950	219
No. 2 MG KD	245	1350	1550	1,700,000	1550 1800	218
No. 2 MG	244	1250	1450	1,600,000	1450 1650	217
No. 2 KD	243	1150	1300	1,500,000	1350 1550	216
No. 2	242	1050	1200	1,400,000	1250 1450	215
No. 3 Dense KD	251	925	1050	1,600,000	1050 1200	224
No. 3 Dense	250	850	975	1,500,000	950 1100	223
No. 3 KD	249	800	925	1,500,000	875 1000	222
No. 3	248	725	825	1,400,000	825 950	221
Construction KD				1,500,000	1150 1300	226
Construction				1,400,000	1050 1200	225
Standard KD				1,500,000	650 750	228
Standard				1,400,000	600 700	227
Utility KD				1,500,000	300 350	230
Utility				1,400,000	275 325	229
Stud KD				1,500,000	875 1000	233
Stud				1,400,000	825 950	232

Note: Terms and abbreviations: Sel. Str. means select structural; KD means kiln dried to a moisture content 15% or less; MG means medium grain; and where KD is not shown the material is dried to a moisture content 19% or less.

(Courtesy of Southern Forest Products Association)

Extreme Fiber in Bending

Extreme fiber in bending, F_b, means that lumber is assessed according to its fiber bending and resisting properties. F_b occurs when loads are applied producing tension and compression as illustrated in Figure 3-3. The tension occurs along the bottom of the member, or the point farthest from the applied load. Compression is induced in fibers along the face nearest to the applied load. The F_b values used in tables such as Table 3-4 have been measured in laboratory tests.

Dense Sel Str Kd
F_b = 2250 psi

Dense Sel Str KD
F_b = 2600 psi

Figure 3-3 Demonstrating compression and tension with load applied and distributed.

Although each grade of lumber has a specific F_b when three or more joists are used to distribute the load in a floor assembly, allowable variations are acceptable. Figure 3-3 shows that the force on a single member must be borne by that member; however, when three (or more) joists are used the load is distributed over a wide surface. In effect this reduces the load on each member and allows for an acceptable variation in the F_b value. Table 3-4 lists both F_b for a single member and F_b with a variation of 1.15 F_b.

Deflection

The two previous topics illustrated certain properties of lumber when under stress. When any load is placed on a member deflection must be considered as well. By definition, deflection is the degree to which a member may safely deflect or bend under load. For appearance sake the deflection must be limited so that sag is not visible. In joist and rafter construction the deflection factor may often be the controlling one in the determination of the size, grade, and species of lumber to be used. For structural joists and rafters it is usually limited to one of three proportions: $S/360$, $S/240$, and $S/180$, where S is the variable of span stated in inches.

Figure 3-4 illustrates deflection. Notice that deflection is the amount of bending a member has when spanning a relatively long distance. This is determined as shown. Where a beam of 16 ft spans an area for its full length, convert 16 ft to 192 in. to obtain the variable S. If a joist calls for a $S/360$ deflection, the total allowable deflection is 1/2 in. The other two ratios, $S/240$ and $S/180$, result in considerably more deflection.

Based on such properties, the building industry has set down the following general guide.

$S/360$ Floor joists in living and sleeping areas. Ceiling joists with attic and plastered ceiling joists with no attic if the live load is predicted 10 psf. Low- and high-slope rafters, plastered ceilings.

$S/240$ Ceiling joists with limited attic, drywall, and a live load of 20 psf. Ceiling joists with no attic, drywall, and a live load of 10 psf. Low- or high-slope rafters, drywall ceiling. Low-slope rafters (3 in. to 12 ft or less) no finished ceilings.

$S/180$ High slope rafters, no finished ceilings, heavy roof covering, loads from 35 to 55 psf. High-slope rafters, no finished ceilings, light roof covering, loads 27 to 47 psf.

Figure 3-4 Demonstrating deflection and the effects of deflection with a variable *S* and different constants 360, 240, and 180.

In summary, floor joists require the greatest stiffness, which means the least amount of deflection. Therefore $S/360$ must be observed. Rafters and ceiling joists may be selected from tables where deflection of $S/240$ is indicated, provided other qualifications as stated before are met. Notice that floor joists are not listed under $S/240$. Finally, $S/180$ may be used for rafter construction only.

Comparative Strength Values

Sometimes it is not easy to choose lumber by its various properties, especially when trying to match species for species. Figure 3-5 provides in graph form a list of many of the common species of lumber used in construction. From the data presented, long leaf pine (which is in relatively short supply) is the best overall species in every category. It would have the best F_b value, the greatest modulus of elasticity, and the least deflection. It would be used for floor joists as well as rafters. Douglas fir ranks only slightly less high in all categories, making it an excellent building material.

In contrast, ponderosa pine is low in bending strength, stiffness, toughness, and hardness. Because of this it is not suited for use as structural members. However, it makes fine moldings and trim and it does have a relatively high nail holding strength.

Dry Lumber—Types and Percentages of Moisture Content

What do the symbols in Figure 3-6 mean? The various symbols KD, MC 15, MC 19, S Dry, S Grn, and A Dry indicate the moisture content of a piece of lumber. They all denote the degree of drying that has taken place. (see p. 55.)

Kiln Dried (KD) means the lumber has been dried in an oven. Stacks of properly spaced lumber are heated in the oven until the moisture content is reduced to 10 percent or less for finish lumber and 19 percent for framing lumber.

Moisture Content (MC). The letters MC followed by a number such as MC 15 and MC 19, simply identify the moisture content of the lumber. It does not identify the method of drying that was used. MC 15 translates to a 15 percent moisture content and MC 19 translates to a 19 percent moisture content.

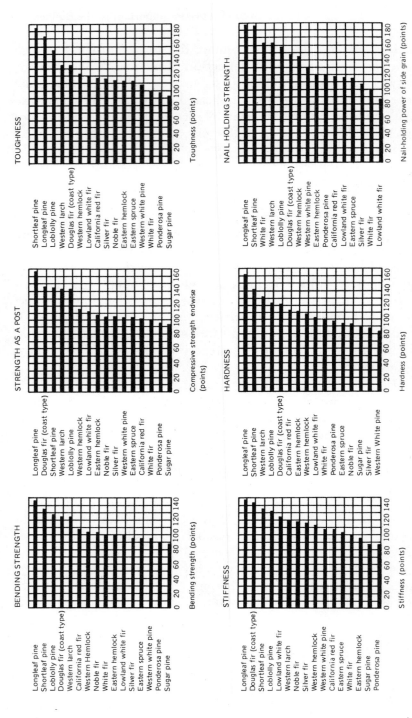

Figure 3-5 Comparative strength values of lumber by species. (*Source:* USDA, Tech, Bulletin No. 408.)

54

Figure 3-6 Six stamps on lumber that reveal its moisture content.

Surface Dry (S-Dry) and Surface Green (S-Grn) indicate a 19 percent moisture content and usually indicate air-dried lumber.

Air dried (A-Dry) is self-defining. It is a process where lumber is properly stacked in a dry climate for periods of three to four months.

Freshly cut lumber may contain as little as 30 percent moisture to as much as 115 percent or more. Once the fiber becomes saturated, the hollow spaces or cells fill with water. Drying the lumber allows the excess water in the cells to evaporate; however, this only reduces the moisture content to an average of 30 percent. Further time or effort must be used to reduce the moisture content in the woody cell walls. This reduction causes fibers to shrink. Thus, unless shrinkage occurs lumber is not considered dried. Naturally, the drier the lumber the more shrinkage has taken place.

This chapter has been developed with a single purpose, to make the reader aware of the many factors that are used in the selection of a piece of lumber for a particular purpose. Many factors were brought out that defined the best lumber suited to trim and non-load-bearing uses, as well as lumber that could be used where stress is a factor as in rafters and joists. The following section examines in problem form the use of the previously developed factors in the selection of a single member in the hopes of bringing into focus their importance in construction.

SELECTING THE PROPER MEMBER

The problems that are to be presented involve all the properties of wood defined in this chapter. Fortunately, the government and many private associations have compiled much of the needed data in tabular form. This simplifies the selection problem. However, two selection methods are presented. Problem 1 uses stated requirements that must be refined from Table 3-4 and tables of species and grades in Appendix A. In contrast, problem 2, which achieves the same results as problem 1, uses prepared tables (Appendix B) rather than calculations to satisfy the requirement needs. Finally, problem 3 shows how another table (Appendix C) may be formed that includes all the requirements of the first two problems and determines in one step the proper structural member to use.

All three problems are faced with the same basic need: Select a species of lumber with the proper grade to support a floor in the living quarters of a residential house. Figure 3-7 illustrates the problem we are faced with. It shows a joist assembly made from 2 X 8 joists that is to span slightly less than 12 ft.

Figure 3-7 Detail of a joist assembly defining span and joist size, but not grade or species.

Problem 1

Requirements:
$S/360 = 1/2$ in.; given $S = 144$ in./360, $S/360 = 0.4$
Modulus of elasticity = 1.9 million psi

F_b including variable factor 1.15 = 1897.5 psi
O.C. spacing 24-in. multiple joist assembly
Live load (LL) = 40 psf; dead load (DL) = 10 psf

Grades available (Table 3-4)	(F_b) psi	(E) psi
Dense Sel Str KD	3050	2 million
Dense Sel Str	2800	1.9 million
Sel Str KD 2400	2400	1.9 million
No. 1 Dense KD	2600	2 million
No. 1 Dense 2350	2350	1.9 million
No. 1 KD and		
appearance KD	2200	1.9 million

Note: All of the above grades exceed the strength (F$_b$) needs for the example floor joist when used in the assembly.

Species based on grade determined above can be found in Appendix A.
Douglas fir–Larch (S-Dry, S-Grn, 19% max. moisture content)
Douglas fir–Larch (North) (S-Dry, S-Grn, 19% max. moisture content)
Southern pine (S-Dry, 19% max. moisture content)
Southern pine (KD, 15% max. moisture content)

The choices available for selection are either douglas fir, larch, or southern pine. No others have the stiffness required. Since both are acceptable, price and availability probably will dictate the selection. Now let's see if the same results can be obtained using different tables.

Problem 2.

Requirements:
$S/360$ = 1/2 in. (same as problem 1)
2 x 8 joists spaced 24-in. o.c. in multijoist assembly
LL = 40 psf; DL = 10 psf
Modulus of elasticity = 1.9 million psi; F_b = 1610
Floor joists 40 psf (LL) and 10 psf (DL); 2 × 8 on 24-in. o.c. (from Table J1 (Figure 3-8 is an extract of Table J1, Appendix B))
Span slightly less than 12 ft.

Tracing across from left in Figure 3-8 at 2 × 8 and 24.0 in. to length of member (11 feet 11 inches), we find F_b = 1610. Trace up from 11-11 to top of the chart and find that E = 1.9 million psi. Therefore, the *grades and species* available are the same as in problem 1 and are obtained from the tables in Appendix A.

JOIST SIZE (IN)	SPACING (IN)	0.4	0.5	0.6	0.7	0.8	0.9	1.0	1.1	1.2	1.3	1.4	1.5	1.6	1.7	1.8	1.9
	12.0	6-9 / 450	7-3 / 520	7-9 / 590	8-2 / 660	8-6 / 720	8-10 / 780	9-2 / 830	9-6 / 890	9-9 / 940	10-0 / 990	10-3 / 1040	10-6 / 1090	10-9 / 1140	10-11 / 1190	11-2 / 1230	11-4 / 1280
	13.7	6-6 / 470	7-0 / 550	7-5 / 620	7-9 / 690	8-2 / 750	8-6 / 810	8-9 / 870	9-1 / 930	9-4 / 980	9-7 / 1040	9-10 / 1090	10-0 / 1140	10-3 / 1190	10-6 / 1240	10-8 / 1290	10-10 / 1340
2×6	16.0	6-2 / 500	6-7 / 580	7-0 / 650	7-5 / 720	7-9 / 790	8-0 / 860	8-4 / 920	8-7 / 980	8-10 / 1040	9-1 / 1090	9-4 / 1150	9-6 / 1200	9-9 / 1250	9-11 / 1310	10-2 / 1360	10-4 / 1410
	19.2	5-9 / 530	6-3 / 610	6-7 / 690	7-0 / 770	7-3 / 840	7-7 / 910	7-10 / 970	8-1 / 1040	8-4 / 1100	8-7 / 1160	8-9 / 1220	9-0 / 1280	9-2 / 1330	9-4 / 1390	9-6 / 1440	9-8 / 1500
	24.0	5-4 / 570	5-9 / 660	6-2 / 750	6-6 / 830	6-9 / 900	7-0 / 980	7-3 / 1050	7-6 / 1120	7-9 / 1190	7-11 / 1250	8-2 / 1310	8-4 / 1380	8-6 / 1440	8-8 / 1500	8-10 / 1550	9-0 / 1610
	32.0					6-2 / 1010	6-5 / 1090	6-7 / 1150	6-10 / 1230	7-0 / 1300	7-3 / 1390	7-5 / 1450	7-7 / 1520	7-9 / 1590	7-11 / 1660	8-0 / 1690	8-2 / 1760
	12.0	8-11 / 450	9-7 / 520	10-2 / 590	10-9 / 660	11-3 / 720	11-8 / 780	12-1 / 830	12-6 / 890	12-10 / 940	13-2 / 990	13-6 / 1040	13-10 / 1090	14-2 / 1140	14-5 / 1190	14-8 / 1230	15-0 / 1280
	13.7	8-6 / 470	9-2 / 550	9-9 / 620	10-3 / 690	10-9 / 750	11-2 / 810	11-7 / 870	11-11 / 930	12-3 / 980	12-7 / 1040	12-11 / 1090	13-3 / 1140	13-6 / 1190	13-10 / 1240	14-1 / 1290	14-4 / 1340
2×8	16.0	8-1 / 500	8-9 / 580	9-3 / 650	9-9 / 720	10-2 / 790	10-7 / 850	11-0 / 920	11-4 / 980	11-8 / 1040	12-0 / 1090	12-3 / 1150	12-7 / 1200	12-10 / 1250	13-1 / 1310	13-4 / 1360	13-7 / 1410
	19.2	7-7 / 530	8-2 / 610	8-9 / 690	9-2 / 770	9-7 / 840	10-0 / 910	10-4 / 970	10-8 / 1040	11-0 / 1100	11-3 / 1160	11-7 / 1220	11-10 / 1280	12-1 / 1330	12-4 / 1390	12-7 / 1440	12-10 / 1500
	24.0	7-1 / 570	7-7 / 660	8-1 / 750	8-6 / 830	8-11 / 900	9-3 / 980	9-7 / 1050	9-11 / 1120	10-2 / 1190	10-6 / 1250	10-9 / 1310	11-0 / 1380	11-3 / 1440	11-5 / 1500	11-8 / 1550	11-11 / 1610
	32.0					8-1 / 990	8-5 / 1080	8-9 / 1170	9-0 / 1230	9-3 / 1300	9-6 / 1370	9-9 / 1450	10-0 / 1520	10-2 / 1570	10-5 / 1650	10-7 / 1700	10-10 / 1790

Modulus of Elasticity, "E", in 1,000,000 psi

Figure 3-8 Table J1 from Appendix B, determining the floor joist to be used by modulus of elasticity, size and length, and F_b.

Problem 3

Requirements: The same as problems 1 and 2.

Refer to Table No. 2 in Appendix C and obtain the grade from the table by:

1. Trace down the left column to 2 × 8, and 24.0 in. o.c.
2. Trace across to 11-11 (11-ft 11-in.) and then up to top.
3. Applicable grades are all those listed in "Grade" columns (1, 2, and 3) and notice that they are the same as those listed in problem 1.

Therefore, because the grades are the same as in problem 1 douglas fir, larch, and southern pine are the species needed for joists. Further tables in Appendix A provide this data.

Summary

What has been proven by the exercise these problems have presented? First, several methods may be used by the builder in selecting the proper species and grade of lumber for a particular job. Next, an understanding of what is included in a table is extremely important if it is to be a working tool. Third, many of the properties of lumber are included in the selection process. Recall that grades are assigned according to limitations and defects. Various drying ratings may be assigned, but in general 15 to 19 percent moisture content is acceptable. Predicted loads must be selected, and the function of the member (joist, rafter, stud) must be known if the proper variable ($S/360$, $S/240$, $S/180$) is to be used. Finally, E and F_b values are needed.

PROBLEMS AND QUESTIONS

1. Define the type of tree that softwood is obtained from. List three types of softwood commonly used in construction.
2. Go to your local lumber yard and make a study of two grades of lumber in stock. Try to list the differences in grain, knots, stains, splits, etc., between them.
3. What are the three select grades of lumber?
4. Of the framing grades which is the best, worst, for framing?
5. What is the allowable difference in fiber-in-bending percentage for *select structure* graded lumber compared with clear straight grain? What is the percentage of modulus of elasticity?
6. Which wood is best for framing?
       a. sapwood   b. heartwood
7. Define specific gravity.
8. Perform the test explained on pages 46 to 47 with a piece of pine and compare your results to those in Table 3-3.
9. What does the modulus of elasticity of a board refer to?
       a. its strength   b. its stiffness
10. What does a board's extreme fiber in bending refer to?
       a. its stiffness   b. its strength
11. If a joist is 14-ft long and is to have a deflection limitation of $S/240$ what would the maximum deflection be?
       a. 1/2 in.   b. 11/16 in.   c. 15/16 in.
12. What does each symbol mean?
       a. KD   b. MC 15   c. S-Grn   d. S-Dry
13. Redo problem 1 (page 00), but change $S/360 = 1/2$ in. to $S/240 = 3/4$ in. and define the lowest grade acceptable as a joist.

4

Energy Efficiency and
the Residential Carpenter

INTRODUCTION

For many years the carpenter has been the person installing insulation. Very early in this century, insulation consisted mainly of dead air, then paper board usually one-half inch thick was installed in walls. Later thin batts of insulation were manufactured and used in walls and ceilings. Only recently has emphasis been placed on full insulation of the residential house.

The most compelling reason for this interest in insulation is the cost incurred in operating heating and cooling systems. Since there is such a significant impact on the homeowner's income because of high utility costs a common demand for energy efficient houses is heard throughout the country. In the northern climate, fully insulated houses reduce heating costs for seven, eight, or more months per year; in contrast, fully insulated houses in southern climates reduce air conditioning costs. These factors, plus increased legislation and improvements in regulations at the federal level have made insulation mandatory.

It therefore follows that a variety of insulating products are available and the carpenter should be able to choose among them. Later in this chapter they are more closely examined, but to generalize there are insulating panels, batts, loose fill, and foam. Insulating also includes proper design and installation of windows and doors, which should include insulation features. The carpenter may also be required to install storm windows and storm doors that trap air between two surfaces and thus have insulating value.

Radiation of thermal energy (measured in BTUs or British Thermal Units) can occur in either direction through a wall or ceiling, door or window, and all too often through the floor. Therefore, special protective measures need to be applied to eliminate the losses or gains of radiation from the ground as well as the air. The carpenter must include various procedures during the construction of the floor assemblies that ensure a control of BTU loss or gain to the house.

Installing insulation is not a very difficult task, even though cramped quarters may be encountered. More significant, though, is the need to understand how insulating boards are currently being used and what factors the use of these boards have on the structural soundness of the building. To illustrate in a general way what could be involved, consider that a 2-in. thick foam insulation panel is being installed over a stud wall. Are let-in corner braces needed to provide adequate strength? It is currently acceptable to use bituminous impregnated fiber board without let-in corners in some areas of the country. Answers to other such questions are provided later in the chapter.

Finally, solar energy is beginning to play a significant role in house construction in many parts of the country. Solar collectors, which include metal, plastic, and water are placed into roof assemblies. What problems do these units pose for the carpenter? He probably needs to learn more about how these systems work, but also needs to know and prepare for the additional weight the roof must carry. It is likely, for instance, that the $S/180$ criteria allowable for rafters cannot be used. Such deflection may prove unsatisfactory, necessitating specialized bracing or trussing.

It should be evident after reading this brief introduction that the carpenter is involved in many ways with establishing an energy efficient house that has sound structural qualities. It is obvious that he needs to have some basic knowledge of thermal control and be aware of the variety of materials he may be required to use.

PRODUCTS AND TERMS

Various products defined in this section are used in residential and commercial construction to provide insulating factors for the control of thermal radiation. Each is assigned an "R" value, which is a numerical value indicating the relative resistance of the product to the penetration or loss of BTUs per hour. All insulations should have a high resistance to crushing and sagging. Some may retain moisture, so care must be used to avoid these problems when using products with limitations.

Cellular-glass insulation board. Boards are available in slabs 2, 3, 4, and 5 in. thick. The assigned R value per inch thickness is 1.8 to 2.2. Thus a 2-in. board would be rated as R-3.6 to R-4.4. This type is classified as rigid insulation, is easily cut to fit, but must be protected from moisture.

Glass fiber with plastic binder. This rigid board is available in thicknesses of 3/4, 1, 1-1/2 and 2 in. and has a resistance value of R-3.3 to R-3.9 per inch thickness. This board is easily cut. If coated with waterproofing it may withstand considerable moisture.

Foamed plastic. Insulation sheets made from polystyrene, polyurethane, and other plastics are usually available in thicknesses of 1/2, 1, 1-1/2 and 2 in. The resistance value may vary from R-3.7 for polystyrenes to over R-6.0 for polyurethane per inch thickness. These panels are currently being used as sheathing in some energy efficient houses.

Dry wall. Dry wall made from paper, asbestos, and other minerals has some ability to control BTU radiation. The R value for 1/2-in. thick sheets is approximately R-0.45. Although relatively low it does contribute to the overall sum of a wall's or ceiling's R value.

Plywood sheathing. All lumber has thermal protection properties. The species of lumber, because of its structure, may have more or less ability to control BTU penetration. However, an approximation of R = 0.156 per 1/8-in. thickness or R = 1.25 per in. is customarily applicable.

Batts. The insulation batt is commonly called a *blanket* and is available in rolls of varying widths and thicknesses. The width does not have any bearing on the R value, only the thickness and type of material used. Batts are available in 1-1/2, 2, 3, 4, and 6 in. thicknesses and are assigned R values ranging from R-3.5 to R-19.

Loose fill insulation. Loose fill insulation is customarily blown into place with a machine although it can be manually placed. It affords a unique insulating quality in that the wool and trapped air combine to form an effective thermal barrier. However, to be effective it usually must be installed quite thickly, 6 in. or more. Its R value ranges from R-19 to R-36.

Trapped-air. Trapped dry air usually called dead air space has an insulating quality if and only if it is stagnated. Such air spaces may be found between sheathing and insulating batts such as stud walls. It may have an R value of R-4 per inch of thickness.

Glass and glazing materials
1. Single strength—R value 1 or below
2. Double strength—R value 1 or below
3. Patio doors—R value 1 or slightly more than one
4. Thermal (vacuum) glass assemblies—R value 1.6 or slightly better.

INSULATING THE FLOORS AND FOUNDATION

In the foundation area, plans and specifications may call for rigid insulation board to be installed. Several typical applications are illustrated in Figure 4-1. In Figure 4-1a and 4-1b the insulation is installed vertically along the inside of the foundation wall. In Figure 4-1c, insulation of the same variety is laid horizontally over the vapor barrier and a short way up the wall, at least to the top of the concrete pour. The use of this material in this manner provides good thermal protection for slabs poured on the ground.

Basement inside walls may also be insulated with rigid insulation or batts. Customarily, 2-in. insulation is used because 2 X 2s are nailed to the concrete block walls on 24-in. centers. Insulation is inserted between the studs. If there has been any previous indication of dampness the walls should first be coated with tar or bituminous material and the rigid board should be used.

The third area that must have insulation is under a joist assembly that is built over a crawl space. There are two methods that may be used to accomplish this objective. Figure 4-2 shows that 1-in. batts are laid on top of the joists and then plywood sheathing is laid on top of it. Figure 4-3 illustrates the second method of insulating a floor joist assembly: batts are installed between the floor joists and are supported by spring-in wood strips, asphalt cement, or friction.

In the first method, flexible blankets of insulation may be used. The strips of wood should be cut approximately 1/2-in. thick and 3/8-in. longer than the space between joists. These strips should be spaced close so that the insulation remains close to the subfloor.

A more rigid batt type of insulation should be used for the cementing or friction method of installation. If cementing the batts in place, an adequate amount of cement must be applied; too little can cause sagging and eventual falling of the insulation. It is

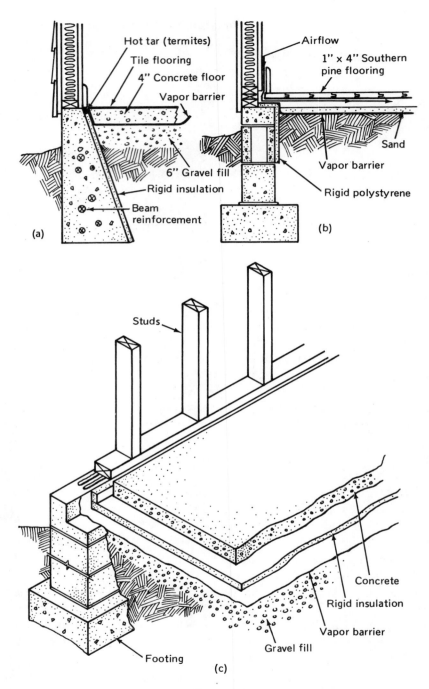

Figure 4-1 Rigid insulation makes good, nondeteriorating thermal protection for foundations, basements, and under slabs.

Figure 4-2 Rolls of insulation beneath subfloors provide minimal thermal protection.

Figure 4-3 Batt or blankets, cemented or compressed between joists provide good thermal protection.

65

also important to stuff pieces of insulation into all areas along the perimeter of the joist wall, as well as around cutouts for plumbing, heating ducts, and electrical wiring.

INSULATING THE WALL

Achieving an R value of R-14 for an insulated wall may be accomplished several ways. It is currently possible and desirable to increase the value beyond R-14 to R-24 by use of rigid polyurethane as sheathing. In the previous section, the average R values were defined for sheathing, air, siding, and batts on a per-inch-thickness basis. The three examples following illustrate how these products work together to accomplish thermal protection.

Single-Wall Construction with Entrapped Air

Figure 4-4 depicts the character of a single-wall system where siding is applied directly to the studs. In this example, an air space of 3/4 inch is planned for and included between insulation and wallboard. This means that if a normal 3-1/2-in. stud is used the insulation thickness is 2- to 3-in. thick. Notice from the legend that each product used in the wall adds to the overall R value of the wall. There is a problem using this insulation method; it is difficult to seal each area between the studs so air is truly trapped. If a leak is present, the air temperature, moisture content, and drafts eliminate the insulating quality. It therefore may be assumed that a better insulated wall may be achieved by substituting R-13 3-1/2-in. batts and eliminating the air space.

Double Wall Construction with No Entrapped Air

In the construction of the double wall as Figure 4-4b shows, sheathing is installed on the outside of studs before siding is applied. In addition, full thickness (3-1/2-in.) insulation is used between sheathing and wallboard. A slightly better total R value is obtained using this method, and the problem of completely sealing off each section to entrap air is eliminated by use of full-thickness insulation.

(a)

(b)

Single wall*	Resistance**		Double wall*	Resistance**
Outside air	= 0.17		Outside air	= 0.17
3/8" Plywood	= 0.47		5/8" Plywood	= 0.78
R-9 Insulation	= 9.00		1/2" Fiberboard sheathing	= 1.32
3/4" Minimum air space	= 3.48		R-11 Insulation	= 11.00
1/2" Gypsum board	= 0.45		1/2" Gypsum wallboard	= 0.45
Inside air	= 0.07		Inside air	= 0.68
	R = 14.25			R = 14.40
	U = 1/R = 0.07			U = 1/R = 0.07

‡R-13 insulation (3-5/8") with or without
foil may be substituted.

* Meets or exceeds
FHA-MPS insulation requirements
for multifamily construction, all regions.

**R values from ASHRAE Guide, Actual
thickness and species groups affect plywood
insulation values.

Figure 4-4 All materials add to the R value of wall, whether single- or double-wall systems.

The 6-in. Wall

The third method of insulating a wall is relatively new, even though the materials have been available for some time. Figure 4-5 illustrates its construction. Sheathing consisting of 2-in. rigid polyurethane sheets are nailed to the outside of the studs creating a 6-in. thermal wall. The addition of this insulation can provide an

R value as high as R-24 to R-26, which is very desirable. The carpenter who builds such a wall must, of course, understand that these rigid panels have little or no strength so they cannot keep the walls in alignment. It therefore becomes absolutely essential that let-in corner braces be installed. Only by inclusion of these braces can proper alignment of walls be established and maintained. In addition to the above condition, this rigid insulation has no holding power, so siding must be installed such that nails or wall ties in brick veneer are securely driven into the studs behind the rigid boards.

R-11 insulation

½" Gypsum board

2" Rigid insulation board

Siding

Figure 4-5 The 6-in. wall.

INSULATING THE CEILING AND ATTIC

The latest trend in insulating ceilings is the "more the better" concept. This is a distinct change from the past, where the theory was that 2- to 4-in. of insulation was sufficient and that the addition of more insulation resulted in very little thermal protection. R values as high as R-36 can be achieved with 12-in. thick insulation. Generally though the depth of ceiling insulation is set at 6 to 6-1/2 in. (R-19) for two reasons: (1) the insulation provides considerable protection in all but the most extreme climates, and (2) ceiling joists are usually 2 × 6 or wider except where trusses are used. Therefore, installation of insulation is relatively simple.

There are several problems that the carpenter may encounter during the construction of attics as shown in Figure 4-6. One is

Figure 4-6 Insulating the ceiling and attic room—air passages must remain intact.

that the attic may be partially finished with rooms framed. In this event insulation is usually installed between ceiling joists as though the attic were unfinished. Another type of problem exists when there is a finished attic. In this event, the ceiling insulation must be installed between the outer wall and partial partition that is the attic room wall, as well as the partition. Thirdly, insulation must be installed between rafters and dormer studs and rafters before dry wall is fastened to these structures.

A word of caution: Almost all cornices have some sort of ventilating screen so that proper air circulation is obtained in the attic area. Improperly placed insulation that blocks the vents and stops the flow of air in the attic could cause serious problems from two points: (1) snow and ice will tend to remain on the roof long after they should have melted away; this puts added strain on all roof components, especially the rafters. Then, too, there is the continual radiation of the temperatures produced by the frozen ice and snow into the attic. (2) In the summer the attic temperature may easily rise to 160 to 180 degrees Fahrenheit. This heat soon penetrates the insulation and ceiling causing higher cooling costs.

All these problems can and should be avoided by properly installing the ceiling and rafter insulation so that air can circulate. The temperature and moisture content of air driven by the wind varies throughout the day and night, alternately cooling and heating the attic.

OPENINGS—WINDOWS AND DOORS

Almost without exception, aluminum and wood window units are manufactured with insulation as a part of the design and construction process. Many years ago a special material, weatherstripping, was used to insulate windows. It was very effective in blocking the passage of air. The insulation included in today's windows is quite effective in controlling the passage of air. Overall, however, a window consists of glass, and the radiation of heat and cold through the glass is considerable as the section on *products* and *terms* illustrated.

The carpenter can do little to change these factors; however, he may be required to install storm windows. If the job is well done a dead air space ranging from 2 to 4 inches is established and this dead air space provides considerable R value thermal protection.

Of course windows may be purchased with thermal panes of glass. Again, this does not affect the carpenter's role. There is, of course, one job the carpenter can do, and that is to stuff insulation between the window frame and surrounding framing prior to finishing the wall. In this way he ensures that no leakage of air occurs around the window.

Doors are the other wall openings that can, if improperly insulated, cause tremendous losses of heat and add problems during summer cooling. Solid doors should be used instead of hollow core and glass panels should be of the thermal variety. The most important and the most difficult part of a door assembly to seal is the space between the door and the jamb or sill. Again, most new door units are weatherstripped with a compression strip around the top and sides. This is generally quite effective in sealing drafts.

The sill presents a different problem. Quite often the threshold must be removed for a variety of reasons; if the door is used over a concrete floor, for example. The installed metal sill is also removed; in other cases no metal sill may have been installed at the factory. A third reason is that the threshold must be built-up so that the door passes freely over thick carpets on the floor. Regardless of the reason, the carpenter is responsible for installing the metal sill so that a completely air-tight surface is obtained. He can install either of two types of metal sill. The compression type (Figure 4-7a) uses a plastic compressible strip to contact the door. The ends of the sill should be puttied and the sill should be caulked prior to fastening it to the floor. The second type of sill (Figure 4-7b) requires that a metal hook strip be nailed to the bottom of the door and this hook slides into a channel on the sill. Putty at the ends and caulking under the sill are required for this installation also.

An extremely effective method of improving the R value at a door area is to properly install a storm door on the outside. This door effectively creates a dead air *space* that has insulating qualities. The door itself may contribute some thermal resistance, especially if it is glazed with a thermal glass.

Compression strip

30 # Felt

Door

Wood threshold (sill)

Compression sill

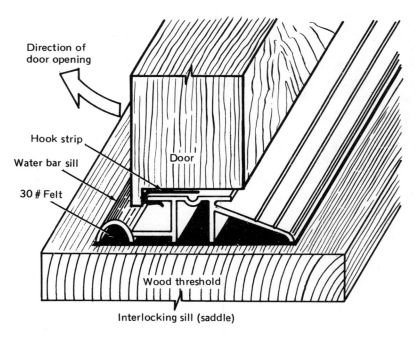

Direction of door opening

Hook strip

Water bar sill

30 # Felt

Door

Wood threshold

Interlocking sill (saddle)

Figure 4-7 Insulating between door bottom and threshold (sill) with metal sills.

72

STRUCTURAL PROBLEMS IN A SOLAR ROOF ASSEMBLY

Solar energy systems are being used more and more by residential and commercial owners. These units are almost always installed on the south-facing side of a roof. In northern climates the roof pitch must be 12-in-12 for best results, whereas in the south lower slopes of 3-in-12 to 7-in-12 may be used.

A solar system consists of panels called collectors, piping made from either plastic or copper, and water. These parts of the system unite in absorbing heat energy from the sun and carry the energy to a storage section in the basement or underground.

The carpenter must construct a more substantial roof than would otherwise be needed. Where, for example, an $S/180$ deflection factor may have been acceptable under normal conditions, the added weight and force created by the solar device prohibits its use. Either of two solutions must be accepted and employed; (1) a different species and grade of rafter must be selected, having an $S/240$ or $S/360$ deflection factor, or (2) trussing between rafter and ceiling must be included where the rafter is reinforced so that the $S/180$ deflection is improved to $S/360$.

Since the solar panels are similar to shallow boxes, sealing their perimeters against rain, snow, and ice is a very real problem. Nondrying mastics, hot bituminous, or flashing must be used. The panels must also be installed in such a manner that rain, for example, has a free path to flow from the roof. Clearly, the carpenter has a vital role in proper contruction of the roof and the solar system being used.

Conclusion

The facts and concepts presented in this chapter have defined the carpenter's role in creating an energy efficient building. He is the one who has the direct responsibility and the opportunity to install insulation, since he is fabricating the building. You should now understand that certain spaces, like the roof, need air passage, and that certain other spaces, under floors and in walls, must be entirely blanketed with insulation. You should also have learned that structural provisions for bracing must be applied when insulation boards having no strength are used as sheathing. During the insulation of doors and windows you should now know that stuffing insulation between jamb and frame is a must and that door sills must be properly installed.

If the carpenter is required to perform these tasks it is because the future owners of the building demand it, or the government, state, or local regulations require it. It therefore is necessary that the carpenter apply his understanding of thermal protection to obtain various thermal resistances (R values).

Broadly stated, Table 4-1 lists R-values for various thicknesses of insulation. These are general values; specific products may vary according to manufacture. And, of course, the other products used in construction add to the overall R value.

TABLE 4-1   Thermal Performance of Insulation by R Value vs.
Thickness Range (in.)

R VALUE	Thickness (in.)	R VALUE	Thickness range (in.)
36+	12–14	13	3 1/2–4
30	10 1/4–12	11	3–3 3/4
24–26	6–7 1/2	8, 9	2 1/2–3
21,22	6–7 1/2	7	2–2 1/4
19	5 1/2–6 1/2	3.5	1–1 1/2
14	5–5 1/2		

Note: Type of insulation is disregarded and values presented are representative of those effective within the thickness span.

PROBLEMS AND QUESTIONS

1. What is the unit used to express thermal energy?
     a. E.E.U.   b. BTU
2. What are the normal areas in a house that are insulated?
3. List two typical places in a house where there is a high loss of heat or air conditioning?
4. If a 4-in. cellular-glass insulating board were used to sheath a house, what would its R value be?
     a. 3.6   b. 8.8   c. 11.0
5. Is glass fiber with plastic binder rated higher in R value than a cellular-glass board?

6. List the total R value for the following wall.
   a. 1/2-in. drywall
   b. 3-1/2-in. batts or blankets
   c. 2-in. glass fiber with plastic binder
   d. outside air
   e. inside air
7. What alterations would you make to construct a wall that had an R value of 18 instead of R-12?
8. Why is it essential that air flow through attic spaces at all times?
9. What makes installing an aluminum threshold or saddle so difficult?
   a. the height of the door over the rugs
   b. weather proofing against air and moisture
   c. screwing it to a concrete floor
10. If a solar panel with water circulating through it is to be installed on the roof will additional bracing or heavier rafters be needed to support the weight?
11. If a roof were originally designed for rafters using $S/180$ rated and solar energy were to be installed, would you still use $S/180$ or _____ instead? Explain.
    a. $S/240$ b. $S/360$

5

Floor Systems

Each and every floor system is an integration of different materials that results in a soundly designed, reliable, and secure assembly. A floor system may require many different types of materials for its construction. Most of these are examined in this chapter. The many grades and sizes of these materials are discussed, along with how to use them effectively. The relationship of a floor system to the blueprint and specifications data is so important that we deal with it at length. It is also important that a floor system be designed properly, whether it be the conventional type or other variety.

DEFINITION OF A FLOOR SYSTEM

The opening sentence of this chapter gives a functional definition of a floor system. Any floor system must be soundly designed. It must be strong and solid, and it must carry all predicted loads. Those loads may be continuous, temporary, or impact types. In each situation the system will normally be engineered to withstand all of these load ratings. It must, of course, be reliable for a very long time, since such an assembly is not likely to be replaced. Then too, it must be properly constructed if it is to have the expected reliability.

However, this definition of a floor system is conceptual. What is needed is a more practical definition, one that relates to its structure or variety of structure. First and probably most important is the need to define the live and dead load limits for residential floor systems. Construction industry standards have been established, setting live loads at 20, 30, and 40 psf (pounds per square foot) for various rooms in a house based upon the purpose of the room. Figure 5-1, a pictorial floor plan of a two-story structure, shows that the living room, family room, and dining room require a live load of 40 psf. These three rooms are typically the *activity* rooms of a house. They are the most lived in, have the greatest concentration of pedestrian traffic, and sustain the greatest continuous, temporary, and impact loads. Contrast the upstairs sleeping rooms and bath, where a 30 psf live load is required. This 25 percent reduction is quite easily understood if the above reasoning is applied. Seldom is there heavy traffic, seldom is there an excessive number of people, and quite likely impact forces are almost nonexistent. Also included in defining the floor system is a common dead load value of 10 psf.

Figure 5-1   Activity areas require 40 psf LL and 10 psf DL. Quiet areas need 30 psf LL and 10 psf DL.

Figure 5-1 also shows floor plans that indicate a need for two systems, the first floor assembly and the second floor assembly. Both are independent and little difficulty should be encountered during their design and construction. However, consider the single-level floor plan shown in Figure 5-2. The same values of 40 psf for activity centers and 30 psf for quiet zone apply; yet the floor system is unified and on one level. What then would be the appropriate approach to take in defining this system?

The most commonly accepted practice is to use 40 psf throughout. Adopting this practice tends to increase the cost of materials slightly. However, the labor costs savings usually more than offset the materials costs. To illustrate, if part of the floor assembly is made from 2 × 8s and part (the quiet areas) could be built with 2 × 6s to meet the 30 psf requirements, adjustments to girder size and height and sill heights would need to be made so that all floor joists would have the same top edge horizontal alignment. Bridging of different sizes would be needed also. The carpenter's tasks related to these adjustments results in considerable labor expenses. Therefore, adoption of a single value of pounds per square foot usually results in the greatest efficiency of resources. As an added thought, the fact that a slight overbuilding in quiet zones is included could be used as a quality selling feature by the builder or salesman.

Figure 5-2 One-story house has different live load requirements for activity centers and quiet zones.

The definition of a floor system also includes the parts of the system that make up the assembly. Each is defined later more completely; however, included are the sills, joists, girders, sheathing (plywood), and underlayment if used. Besides materials there are the construction techniques that include the spacing of joists, the type of sill and girder construction selected and incorporated, and the reinforcements needed to integrate the materials.

The definition of a floor system immediately creates a need to examine the parts used to form it.

PARTS OF THE SYSTEM

Girders

It is customary that a girder be installed either within the joists spanning the width or length of a floor or under the joists. Methods A and B shown in Figure 5-3 are used where the top surface of the joist and girder must align. The illustration of method A shows that each joist must be matched to fit over the support ledger. Employing this technique does not have any adverse effects on the girder, but does reduce the bearing width of the joist.

Contrast method B with A and note that the full joist rests on the ledger. However, this method may cause a girder to be built that exceeds the structural strength and stiffness requirements necessary. Its use also creates a 2-in. or larger offset below the joist.

Method C shows a simple way of including a girder where it may be cut or built-up to the required strength and stiffness without concern for the size of the joists used for the floor. It may be constructed from a solid member or built-up as shown. Although the girder is quite heavy, it is relatively easy to install. The joists will rest on top of this girder and are toenailed to it.

Girders may be made of any size lumber, but the usual sizes are: 4×6, 4×8, 4×10, 4×12, 6×6, 6×8, 6×10, 6×12, and 8×8 or 8×10. The most common sizes are the 4×6, 4×8, 4×10, 6×8, and 6×10. These are large members, but they must carry a larger load for many years. One way to reduce the needed size is to limit a girder to the shortest length possible.

In addition to the varieties of lumber sizes, girders may be purchased in the form of steel I-beams and WF (wide flange) beams; both are shown in Figure 5-4.

Figure 5-3 Three methods of using a girder; in-line, in-line full joist support, and under joists.

Figure 5-4 Steel girders (beams) I and wide flange (WF) varieties.

Floor plans and sometimes foundation plans state the size and placement of girders. It is also possible that the girder size is listed in the specifications or footnotes, or that the carpenter must make the decision about the girder's size and placement. Thus the selection of a girder must be made on the task it must perform. This single member has more weight and stress placed upon it than almost any other part of a building.

A girder spans an open area, with each end resting on a solid foundation. It may be that the girder spans the entire building, but the overall span is broken down into several smaller spans by using columns or piers. This procedure is extremely important, as will be shown later. But briefly, the total weight on the girder is reduced greatly if the spans are held to 8, 10, or 12 ft.

A girder must also carry the weight of part of the floor resting on it or supported by it. Figure 5-5 illustrates this. The building shown is 24 ft wide by 32 ft long. A girder is built to run 32 ft midway through the building. It is supported on the outside walls and by three piers or columns. The result is a clear span of girder just under 8 ft. Since the joists are 12-ft long either side of the girder, one-half of their length is supported by the outside wall. The other half (6 ft) is supported by the girder. In this example the girder supports 6 ft of joists each side of itself for a total of 12 ft. Notice that the amount of support required of the girder is equal to one-half the building's width. This condition exists everywhere a single girder is installed between outer walls. The shaded area in Figure 5-5 represents a potential 50 psf load.

Figure 5-5   Floor area supported by a girder(s).

Figure 5-6 represents just section A of the shaded area of Figure 5-5. The first pier is set 8 ft from the end wall and sets the span of the girder at this distance. Since the S distance is 12 ft, a grid is made of 1-ft squares, each having a combined load of 50 psf. This means that each 1-ft run of the girder over its clear span must support 600 lb. In the 8-ft span the total weight on the girder is 4800 lb. This is considerable and would indicate the use of a 4 × 8 or 6 × 6 girder made of No. 2 grade lumber and possibly some No. 3 grade species of sufficient strength.

This is not all there is to selecting the proper size girder. Suppose there is a load-bearing wall placed on top of the girder. This means that there will also be ceiling joists or rafters that bear on the partition. Figure 5-7 shows a set of conditions where there is a bearing partition, ceiling joists, and an unfinished attic space. In this situation the girder must support the weight across the S distance of the ceiling joists and drywall or other materials, plus the partition and all live loads *in addition* to the first floor. Notice how quickly the different loads add to the total load the girder must carry.

Figure 5-6 Defining the load on a girder using a 1 ft² grid layout.

S = LL & DL of floor bearing on girder

Figure 5-7 Defining the loads a girder must bear.

Rafter

20 # LL

Attic

10 # DL
Ceiling
Joists and drywall

10 # DL bearing partition

40 # LL floor

10 # DL

90 # /ft load on girder

Figure 5-8 Adding up the loads a girder must carry; attic, framing, walls, etc.

Figure 5-8 shows in more detail how the loads of each part contribute to the whole load. Beginning with the attic, a live load of 20 psf is anticipated primarily from storage of personal belongings. Next a 10 psf dead load for ceiling joists and ceiling materials is added. The load-bearing partition and its wall coverings add another 10 psf to the load on the girder. The first floor is designed for a 40 psf live load and 10 psf dead load. This brings the total load on the girder to 90 psf. To imagine the effect this has, consider Figure 5-6 again; but instead of each square foot having 50 lb, each has 90 lb. The same length of girder would need to carry almost twice as much weight. This would mean that a girder made of a higher quality (grade) and species of lumber or a larger size would be needed.

Several tables for the selection of girders are provided in Appendix D.

Beams

A beam is a member that is usually larger than a joist. Although a beam may be used as a joist on, say, 24-in. centers, they are usually spaced farther apart. The beam and plank floor system is a common example of the use of a beam. As with joists, the beam is measured in nominal widths and thicknesses and is rated for stiffness (E) and strength (F_b) much like a joist. This means that live and dead loads similar to those on living areas, quiet areas, and even roofs may be applied. Several examples of beams of different sizes are shown and explained in the following figures so that some understanding of how the size and length of a beam play a significant role in the grade selected. These examples represent floor beams where a stiffness of $S/360$ is needed.

The stiffness or *modulus of elasticity, E,* and strength or *fiber in bending, F_b,* are the two characteristics necessary to determine the proper grade of lumber that should be used as a beam. Table 5-1 is developed for a nominal 4-in. by 6-in. beam so that the reader may begin to make a comparison of the relative usefulness of the various sizes.

First note in the table's title that the conditions of "40 psf LL and 10 psf DL" are the initial parameters selected. The reason for selecting these values reverts back to the fact that all activity areas in a house need these values. A live load of 40 psf and a dead load of 10 psf are established by most of the local and national building codes. The modulus of elasticity is developed from the live load value only, in this case 40 psf. The fiber in bending is a combined value of the live load (40 psf) plus dead load (10 psf) resulting in a combined value of 50 psf.

The first observation made from the tabulated data is that this size beam will adequately support any span up to 8 ft made from any grade of construction lumber provided its E equals or exceeds 280,000 psi and F_b equals or exceeds 275 psi. Figure 5-9 shows the girder and lists all the minimum grades acceptable for each length. A beam of *standard* may be used for a 4 × 6 beam spanning 8 ft. It must have an E and F_b as stated before.

TABLE 5-1   4-in. × 6-in. Beam; Strength and Stiffness Requirements (40 psf LL; 10 psf DL)

					Length of Span (Ft)				
	8	9	10	11	12	13	14	15	16
$(E)^{a,b}$	0.28	0.41	0.56	0.74	0.96	1.25	1.51	1.9	—
$(F_b)^b$	275	345	430	510	610	720	840	960	1090

[a]Multiply value in table by 1 million to obtain stiffness (modulus of elasticity).

[b]E and F_b are both measured in pounds per square inch (psi).

If the span of the beam is 13 ft, for example, Figure 5-9 lists a grade No. 3, No. 2, or No. 1 as appropriate. Table 5-1 lists an E of 1.25×10^6 or 1,250,000 psi, and F_b of 720 psi. A search through table W-1 in Appendix A, which is designed primarily for joists and rafters but is accurate for beams as well, provides the following:

Grade No. 3	Grade No. 2	Grade No. 1
Douglas fir–larch	Douglas fir– larch	Eastern hemlock
Douglas fir (north)	Douglas fir (north)	Eastern hemlock (north)
Southern pine	Douglas fir (south or hem-fir)	Eastern spruce
	Hem-fir (north)	Engleman spruce
	Southern pine	Alpine fir

The listing shows that ten species of lumber qualify, and of those only three may be used with a grade No. 3. All other species must be of a better than No. 3 grade. Also note that the grade for the eastern woods must be No. 1.

Figure 5-9   4 x 6 beam—40 psf LL, 10 psf DL, span and grade to use for various spans.

Finally, the 4 × 6 beam is effectively used with spans of 5, 6, 7, up to 14 ft, and in some cases 15 ft. When one of 15 ft is used, Douglas fir larch or southern pine species must be used. It might be better to select the next larger girder size if a span of 14 ft is called for.

Table 5-2 and Figure 5-10 combine to provide a picture of the requirements for a 4-in. by 8-in. beam of various lengths. The dimensions of this beam provide considerable strength to a floor assembly.

TABLE 5-2   4-in. x 8-in. Beam; Strength and Stiffness Requirements
(40 psf LL; 10 psf DL)

					Length of Span (Ft)				
10	11	12	13	14	15	16	18	20	22

	10	11	12	13	14	15	16	18	20	22
E^a	0.25	0.32	0.42	0.54	0.68	0.8	1.0	1.4	1.95	—
F_b	245	295	350	410	460	550	625	800	980	1200

[a]Multiply value by 1 million.

Figure 5-10 4 x 8 beam—40 psf LL, 10 psf DL, span and grade to use for various spans.

Table 5-2 shows that this beam may be used for any short spans; but only when spans of 10 ft or more are expected do stiffness and strength values become important in the selection of species and grade. The table indicates that the stiffness values are within various grades up to 18 ft, where a modulus of elasticity of 1.4 million psi is needed. However, if an exceptionally fine grade of Dense Sel Str douglas fir larch or southern pine is available, a 20-ft beam may be used. Thus far the tabular data has illustrated that strength requirements are most easily met.

In every case the member's strength exceeds the minimum required pounds per square inch. Not so with the stiffness value E. The upper limit is reached long before the strength value is maximum. Figure 5-10 illustrates that in almost all spans the beam may be made from No. 3 graded members. Careful selection of species within these assigned grades is necessary beginning with spans of 15 ft and longer. Although the fiber in bending values are relatively small, the modulus of elasticity values rise rapidly.

The 4 X 10 beam shown in Figure 5-11 and detailed in Table 5-3 is one generally used where extremely long spans are a design feature. For spans below 16 ft this size beam would be very expensive, especially when several other types such as the 4 X 8 or 3 X 10 would provide adequate stiffness and strength. As Table 5-3 indicates, beams up to 26 ft are well within the stiffness and strength parameters of the various grades of lumber available today.

A review of Appendix A shows that the selection by species narrows quickly when beams of 25 ft and 26 ft are used. For example, only the southern pine (KD) in Dense Select Structural meets the modulus of elasticity requirements needed for the 26-ft beam. However, the Douglas fir larch, the Douglas fir (north) and the southern pine are adequate for the 25-ft beam and the shorter lengths.

The data on beams that have been presented thus far are restricted to a 40 psf LL and 10 psf DL. Therefore the beams would not be reliable if loads above these were required.[1] It also follows that beams cut on these specifications would exceed the minimum requirements if loads less than those listed were designed into the structure.

---

[1] Complete tables are published in many forms for beams by various construction industry and lumber associations. See reference list on page 373.

Figure 5-11   4 × 10 beam 40 psf LL, 10 psf DL, span and grade to use for various spans.

TABLE 5-3   4-in. × 10-in. Beam; Strength and Stiffness Requirements
(40 psf LL; 10 psf DL)

	Length of Span (Ft)						
	16	18	20	22	24	25	26
E^a	0.48	0.7	0.95	1.25	1.6	1.85	2.0
F_b	380	480	600	720	860	950	1100

[a]Multiply value by 1 million.

TABLE 5-4   4-in. × 8-in. Beam; Strength and Stiffness Requirements
(30 psf LL; 10 psf DL)

	Length of Span (Ft)								
	10	11	12	13	14	15	16	18	20
E^a	0.2	0.24	0.31	0.4	0.5	0.62	0.75	1.05	1.4
F_b	200	240	280	330	380	440	500	640	800

[a]Multiply value by 1 million.

Figure 5-12 4 × 8 beam 30 psf LL 10, psf DL, span and grade to use for
various spans.

Table 5-4 and Figure 5-12 combine to provide data on a 4 × 8 beam that has stiffness and strength requirements of 30 psf LL and 10 psf DL. Reducing the load on a 4 × 8 beam means that less demand is placed on it. Therefore, the beam can span lengths more easily, and lower quality lumber plus greater varieties of acceptable species make it easier to obtain. Table 5-4 shows that spans less than 10 ft may be made from any construction grade since the E value is less than 200,000 psi and the F_b value is only 200 psi. At the other extreme of the table, a beam 20 ft long only requires an E of 1.4 million psi and F_b of 800 psi. This means that a No. 2 or No. 1 grade lumber may be used. Again, this does not apply to certain species.

Figure 5-13 4 x 10 beam 30 psf LL 10 psf DL, spans and grade to use for various spans.

TABLE 5-5   4-in. × 10-in. Beam; Strength and Stiffness Requirements
(30 psf LL; 10 psf DL)

	Length of Span (Ft)									
	13	14	15	16	18	20	22	24	25	26
E^a	0.2	0.24	0.3	0.36	0.52	0.7	0.94	1.2	1.4	1.55
F_b	205	240	270	310	390	480	580	700	750	810

[a]Multiply value by 1 million.

The fifth and final example of beam characteristics is shown in Figure 5-13 and detailed in Table 5-5. This is a 4 × 10 beam that will support a 30 psf LL and 10 psf DL. This beam would be included in floor design and construction in quiet areas of a house, most probably in the split level and two-story house. Lumber graded No. 3 should be used for all lengths up to 22 ft. Lengths of 24, 25, and 26 ft require a closer examination of E and F_b within grades and species to ensure adequate stiffness and strength.

Anchoring Girders or Beams

A girder or beam must span the width or length of a building or at least a large segment of the floor area. It therefore must be anchored at both ends. Figure 5-14 shows several typical methods. If installed between masonry units, termite shielding must be used. If inserted or made part of the joist assembly, anchoring is made to header joist and L-type sill. A sound installation means that (1) the girder rests on a minimum surface area of 1-1/2 inches wide, (2) it is properly protected against decay and infestation from termites, and (3) it is soundly anchored to prevent shifting and twisting.

Sills and Types of Sills

In earlier days of construction the member that was attached to the foundation was commonly called a *bed plate*. This term has given way to the now common *sill*. However the term sill is just as frequently referred to as an assembly consisting of a flat member resting on a foundation plus a vertical member usually the size of a

Figure 5-14 Installing girder (or beam) on a bearing surface; anchoring and adequate seating.

joist. Where such combinations of members are used, the assembly is called an L-type sill. There is also a box-sill type that is sometimes used. It is shown and explained later.

L-Type Sill. Figure 5-15a shows a built-up L sill. This arrangement of horizontal sill (plate) and beam or header is the one most commonly used by carpenters. It is easily assembled and reliable. It provides ample surface for joist end support and makes vertical alignment of joists a relatively simple task. Notice the symbol L drawn over the joist. The combination of sole plate and header form the letter L and hence the name L sill. Subflooring, such as plywood, is usually laid flush with the outer edge over the joist and sill assembly. Walls are erected on top of the plywood or other sheathing. This method of framing a sill does not create a fire stop, but when the sheathing and wall sill are installed a fire trap is constructed. Flames in the basement area cannot pass into the wall until they first burn through the subflooring and wall sole plate.

Box-Type Sill. The box sill assembly is a variation of the L sill and derives its name from the concept of a box with a bottom side and top. Figure 5-15b shows a box sill assembly. Note that it is built up from the sill header-joist and sole plate of the wall. The sole plate of the wall is nailed directly to the top of the header and each joist. Subflooring is then cut to fit against the wall's sole plate.

Figure 5-15 L and Box type sills.

Installation of Sills. Most experienced masons can build a wall that meets most specifications for plumb and level. The specifications do, however, allow for varying degrees of error. The one of most concern to the carpenter is the level accuracy of the wall's top course. Masons are allowed 1/2 in. from level in each 20-ft run. It would be nice if their accuracy were 1/2 in. from corner to corner, but there are no guarantees for the carpenter that the mason will achieve this. Therefore, the carpenter must be prepared to install a sill that is level, but must fully expect the masonry wall to be unlevel.

Two tasks are accomplished simultaneously when a level sill is installed. First, a new standard of level is established for the house to be built upon. Second, the grout or mortar bed used to fill in irregularities between sill and foundation seals the entire length and adds both an insulation factor as well as reduces the opportunity for insect infestation.

The mason or his laborer should prepare a batch of mortar so that after sills have been precut and drilled and are ready for installation, they can apply sufficient mortar along the wall. Bolting the sills should spread the mortar and excesses should be troweled away.

If the wall's level is exceptionally good, caulking can be used in place of mortar. It can be placed in ribbons along the wall and it will seal the spaces between foundation and sill.

Joists

Following the construction of the sill, the joists are cut and installed. The selection of joists is, of course, dependent upon the span they are to cover and their live and dead load requirements. Appendixes B and C provide data for various size joists with on-center spacing as well as grades to use for numerous spans.

Recall that living or activity areas within the house need 40 psf live load and 10 psf dead load characteristics. Therefore, selection of the joist size and grade must be taken from the tables having these values. Also recall that sleeping quarters may have their floors framed from members that properly carry a 30 psf live load and 10 psf dead load.

Selection of a joist must include *stiffness*. Recall that stiffness is a measure of the modulus of elasticity, and *strength* is a measure of fiber in bending. If a floor is to remain sound for many years, both qualities must be possessed by each joist and must be properly defined. The stiffness value, E, must be sufficient so that a floor can deflect and return to its original condition time after time. Any member included in a floor assembly that cannot meet these requirements degrades the entire assembly.

Having the qualities included in each joist is important, but the carpenter must take the opportunity to make use of the stiffness and strength. For example, each member that is milled as a joist has a *crown side*. This crown side indicates *the area to be compressed*. The concave underside should be the one to be put into tension. The crown side is generally placed under pressure by live and dead loads that tends to nullify some of its tension by compression. The underside is already compressed and the loads stretch the compressed fibers releasing some of the compression.

Consider now what problems can occur if one or more joists are placed crown side down and the remaining ones are placed crown side up. Figure 5-16 shows that the loads cause the crown-up joists to deflect downward and recover when the load is removed. However, the joist with crown down is already deflected and cannot easily deflect more. So what happens? Uneven movement causes unequal strain and stress on subflooring and finished flooring; squeaks result and sagging often accompanies the squeaking. Worst of all, people walking across the floor get the sensation of the floor giving way. Details in Figure 5-16 are, of course, somewhat exaggerated so that the concept can be readily seen. Separation of flooring from joist rarely exceeds one-quarter inch at the most negative (lowest) point in the span. Nevertheless, careful alignment of joist is absolutely essential to sound construction.

Figure 5-16 Improper installation of joist (crown) resulting defective floor assembly.

Load-bearing wall
or beam

Stairway opening

Width

Length

Tail joist

Double trimmer
joists

Joist

Double
header

Wall or foundation

(a)

Double trimmer

Joist

Stair

Beam or wall

Joist hanger

10' 0" Max. header length
when designed as beam

Double header

Double trimmer

Tail joist

Post or other support for
header and trimmer

(b)

Figure 5-17 Framing for floor openings—headers, tail joists, and trimmer joists.

Headers, Tail Joists, and Trimmer Joists

Headers are frequently needed to complete the framing of an opening in the joist assembly. The primary functions of headers are to support the *tail joists* and to define the width or length of the opening. Frequently a bearing wall is placed across headers near stair wells, for example. This condition would make it extremely important that the headers be doubled and supported if necessary by hangers or support beneath the header or well nailed to other floor members as Figure 5-17 illustrates.

Both drawings in Figure 5-17 include headers; notice that in Figure 5-17a the headers are short members that span the width of the openings. However, in Figure 5-17b they are relatively long and span the length of the openings. These two illustrations help define the word *header*. Another way to identify a header is to notice that tail joists are always nailed to them.

Tail Joists are pieces of regular joist stock that are cut to fit from header to outer wall or girder or some form of bearing surface. Tail joists must be well nailed to the header. Customarily, three 16d common nails are used to secure the tail of the joist to the header. Although not shown, the tail joists can be made with a cutout so that they will rest on a ledger board attached to the lower vertical edge of the header. (It would be similar to the ledger shown in Figure 5-3).

Trimmer Joists are the third and final member of framed opening in a floor assembly. These members are regular joist stock cut to fit between supporting surfaces and nailed soundly to the joists that support the headers. In Figure 5-17a the trimmer is nailed either side of the full joist that the headers are nailed to, and run perpendicular to the header. In Figure 5-17b the trimmers also run perpendicular to the headers and parallel to the floor joist run.

Support Joist for Load-Bearing Wall (Parallel to Joist Run)

There is often a need to have bearing walls run parallel to the run of floor joists. In customary framing and design methods these walls may be placed over a standard on-center spacing joist as Figure 5-18a shows; however, many times the wall is located between joists. Since the wall is load bearing, it means that ceiling joists and maybe some portions of the roof's force may be concentrated on the wall. If no additional support is added to the floor assembly the sheathing or subflooring must carry this load, and it is not strong enough to do so. Therefore, another joist must be placed between two normally spaced joists so that it is centered under the bearing wall (Figure 5-18b). In this manner there is full support for the full length of the wall.

Figure 5-18 Load-bearing joists perpendicular to run of joists.

If a MOD 24 or a 24-inch system of design were used instead of the conventional construction method, all bearing walls would fall over normal on-center spacings of joists and the need for the additional joist is eliminated. This means that each bearing wall would be placed as shown in Figure 5-18a.

Figure 5-19 Non-load-bearing wall supported by blocking in joist assembly.

Support Framing for Non-Load-Bearing Wall (Parallel to Joist Run)

Since the wall is not carrying a load such as ceiling joists or roof it is classified as non-load-bearing. This means that support within the floor assembly need only be designed to carry the dead load of the wall. The customary method of providing support where the wall is located between on-center spacings of floor joists is shown in Figure 5-19. Nominal 2 X 4 or 2 X 6 members frequently referred to as "cats," "solid bridging," or "blocking" are securely nailed between joists on 3- to 4-ft centers.

Bridging

Industry standards prescribe the use of bridging where the span of a joist exceeds 8 ft. A row of bridging made from nominal 5/4 x 3-in. stock, joist stock or metal straps should be nailed midway between the end supports of the joist where spans are 16 ft or less. Where spans exceed 16 ft, two rows of bridging must be used to comply with the standard.

There may be, on the part of some builders, a desire to omit bridging because of the labor and material costs. Should such a plan be followed, the result would be a floor assembly that gives with concentrated loads. The function of bridging is to cause a concentrated load on the floor assembly to be distributed throughout the joists nearby.

The downward pressure caused by the weight is not absorbed by a single joist or two adajcent joists. Rather the energy of the pressure is transferred laterally into bridging and thereby to adjacent and remotely adjacent joists. True the subflooring also aids in distributing the force, but bridging is vitally important. By distribution of the pressure as described the joists do not deflect to their maximum and so avoidance of damaging stress is achieved.

Subflooring

The most commonly used material for subflooring is plywood. It is a unique material possessing many properties that allow a variety of construction techniques to be applied. It is available in many sizes, from the standard 4 X 8-ft sheet to those ranging up to 16 ft long. It is also available in many different thicknesses, grades, and types. All of these properties make the carpenter's job somewhat more difficult rather than simplifying it.

As each floor system is designed, the selection of plywood for a subfloor must meet all necessary requirements. Or the opposite view may be taken, which is that the selection of plywood may in large part dictate the construction method of the joist assembly. Several adaptations are shown in the next section of this chapter. For the moment, though, the characteristics of plywood available for subflooring are illustrated in Table 5-6 and are examined.

TABLE 5-6 Identification Indexes for Plywood Grades
(Courtesy of American Plywood Association)

Thickness	C-C Exterior C-D Interior			Structural 1 C-C and C-D Structural II[b] C-C and C-D	Structural II C-C and C-D
	Group 1 *Group 2*[a]	*Group 2 or 3* *Group 4*[a]	*Group 4*	*Group 1*	*Group 2 or 3*
5/16 in.	20/0	16/0	12/0	20/0	16/0
3/8 in.	24/0	20/0	16/0	24/0	20/0
1/2 in.	32/16	24/0	24/0	32/16	24/0
5/8 in.	42/20	32/16	30/12	42/20	32/16
3/4 in.	48/24	42/20	36/16	48/24	42/20
7/8 in.[c]	—	48/24	42/20	—	48/24

[a]Panels conforming to special thickness provisions and panel constructions of Paragraph 3.8.6. of PS 1.

[b]Panels manufactured with Group 1 faces but classified as Structural II by reason of Group 2 or Group 3 inner plies.

[c]Panels thicker than 7/8 in. shall be identified by group number.

The most widely recommended types of plywood used as conventional subflooring are: Structural I and II CC, Structural I and II CD, CD Interior Sheathing, CC Exterior and (not shown in Table 5-6) 2-4-1 Interior. The interior types should be used where moisture is not expected to be a problem. Such places as subfloor over a basement area or between first and second floors are appropriate for this type of plywood. Floor assemblies over crawl spaces and those exposed to humid or wet areas must have the subflooring made with exterior glues, so exterior types are best suited.

The letters CC and CD denote the surface condition of the plywood. The first letter denotes the best surface and in C grade small knots are allowed as well as discolorations, stains, and some splits. The surface may be unsanded or touch sanded. The back surface in CC is the same as the front. In CD more defects are permitted; larger knots are allowed and more severe splits may be included.

The groups 1, 2, 3, and 4 shown below refer to group species of woods that are used to construct the plywood. Group 1 species are the best types and provide the greatest strength and stiffness. Group 2 species also contain many fine types that make strong and stiff plywood but are slightly less strong and stiff. Groups 3 and 4 contain the remainder of the species, many of which are found to have a low modulus of elasticity and fiber in bending.

Group 1:	Group 2:	Meranti	Spruce, black
Birch, sweet	Cedar, Port	Pine, pond	Spruce, red
Birch, yellow	Orford	Pine, red	Spruce, white
Douglas fir 1	Fir, California	Pine, western	
Larch, western	red	pitch	Group 4:
Maple, sugar	Fir, grand	Pine, white	Aspen, bigtooth
Oak, tanbark	Fir, noble	Spruce, sitka	Aspen, quaking
(tanoak)	Fir, Pacific	Sweetgum	Birch, paper
Pine, Loblolly	silver	Tamarack	(canoe)
(swamp)	Fir, white		Cedar, incense
Pine, longleaf	Hemlock,	Group 3:	Cedar, western red
(Georgia)	western	Alder, red	Fir, subalpine
Pine, shortleaf	Lauan, almon	Cedar, Alaska	Hemlock, eastern
Pine, slash	Lauan, Bagtikan	Jackpine (gray)	Pine, eastern white
(Caribbean)	Lauan, red	Pine, lodgepole	Pine, sugar
Pine, southern[a]	Lauan, tangile	Pine, ponderosa	Poplar, western
	Lauan, white	Pine, spruce	Spruce, Engelmann
	Maple, black	Redwood	
	Menkulang		

[a]"Southern pine" is sometimes used to designate all the pine trees listed here, including the Carribean pine since it grows in Florida, as well as in Cuba and the Bahamas.

Finally, a selection of thickness must be made by using the data in Table 5-6. Each number printed under a group heading is composed of two numbers separated by a slash. The left number is used to indicate on-center spacing of rafters and the right number indicates on-center joist separation. Close examination of the table shows that the minimum thickness of plywood usable for subflooring is 1/2 inch in group 1 and some group 2 types, in several grades. No group 3 or 4 types are allowed. The data also show that if joists are spaced 24-in. o.c., 3/4-in. plywood is used from group 1 types regardless of grade.

In addition to the selection process needed, a special nailing sequence is important and should conform to 8-in. spacing on intermediate joists and 6-in. spacing around the perimeter of the sheet. Expansion provisions of 1/16 in. must be included in high humidity and otherwise damp areas; spacing of the plywood sheets is easily accomplished by using 4D finish nails as separators.

Several types of plywood are made tongue and groove. Where this type is used no supports need be included between joists where plywood edges join (Figure 5-20).

However, where square-edge types of plywood are used and no underlayment plywood is scheduled for use, blocking, as shown

Figure 5-20 Blocking to provide edge support for plywood sheathing.

in Figure 5-20, provides nailing as well as adhesive bonding surface. This blocking is made from nominal 2 × 4 or 2 × 6 stock, and it may be installed flat surface up providing ample bonding and nailing surface for securing the plywood. It may also be installed on edge where a 3/4 in. surface is available for each plywood sheet to be secured to. The selection of method is arbitrary. Consider the possibility that bridging is necessary; solid blocking and the span of joist is close to 16 ft. A row placed 8 ft from the outside wall or on center could perform a double function. It could be the blocking as well as the edge nailing surface for the plywood sheathing.

PLANNING THE FLOOR ASSEMBLIES

The preceding pages have provided the foundation needed for planning the construction of a floor assembly. A simple one-and-a-half-story house has been selected for the planning. The upstairs is a finished attic with four bedrooms where the ceiling follows the rafter line part way and the walls are built inward several feet from the outer walls, front and rear. The downstairs floor must be designed for activity and rest areas, and is to be built over a block foundation. Figure 5-21 shows the exterior view. Notice that the gable roof has gable ends on the sides of the building; this means the rafters run front to back. If the floor and ceiling joists run front to back construction will be easy. This alignment of joists and rafters is not absolute but it is the rule rather than the exception.

Figure 5-21 Framed house with expanded finished attic.

Criteria

1. Design the first floor assembly so that when constructed it will support a 40 psf LL and 10 psf DL.
2. Design the second floor assembly so that when constructed it will support a 30 psf LL and 10 psf DL.
3. Include provisions for supporting non-load-bearing walls that run parallel to the joists.
4. Construct a girder that will carry the load and span 11 ft 6 in. for the first floor.
5. Construct a girder to be flush with joist edges for the second-floor assembly to span the dining room, and the kitchen or living room.
6. Ensure that load-bearing partitions are aligned over first-floor girder(s) as closely as practical.
7. Install bridging as required.
8. Frame the stairwell opening in the second-floor assembly.

First-Floor Assembly Plans

Figure 5-22 shows the floor plan for both the first and second floors. Plans *A*, *B*, and *C* of the figure show possible designs of the floor assemblies. Each design differs as is explained.

In plan *A* the girder is placed off center of the 24-ft span such that it is centered 14 ft back from the front of the house. This places it directly under the short kitchen wall, and the hall wall leading into the bathroom and back bedroom. The partitions that will rest on top of the girder are load bearing, so this indicates a proper position for the girder. Notice, however, that the partition between bedroom closets is centered in the building and is off the girder. This does not present a serious problem because the length of the partition is roughly 6 ft and only 3 to 4 ceiling joists will bear on this wall.

The piers that will support the girder should be placed 8 ft o.c. so that the girder spans equal distances and the load is uniform along its length. This allows for a selection of a 4 × 8 or 4 × 10 girder.

If the girder is placed under the floor joists, as *method C* of Figure 5-3 shows, a 4 × 8 girder is adequate. Since the floor assembly is built on a foundation (over a crawl space or basement)

Figure 5-22 Designing the floor assembly for first and second floor.

107

such an arrangement is possible and practical. If used, the floor joists will rest on top of the girder and will be nailed to the girder.

If for some reason the joist and girder must have flush surfaces, a restriction is imposed on the size of girder. The longest span of joist is just under 14 ft (from outer wall to girder center). This means that the joist used for the floor must be either a 2 X 10 if 24-in. o.c. spacings are used or a 2 X 8 if 16-in. o.c. spacings are used. A careful look at Appendix C shows that the best (Dense Sel Str) quality 2 X 8s must be used. But a lower quality (No. 2) 2 X 10 can be used even though on-center spacings are further apart. This means that the girder must be made from either 2 X 8s for 2 X 8 joists or 2 X 10s for 2 X 10 joists.

The carpenter must make a trade-off since a variety of possibilities exist. Clearly, though, the girder placed as shown in plan *A* is ideally located. It provides proper support for floor joists as well as load-bearing walls.

The balance of construction techniques such as L-type sill, nailing schedule for joists, and a row of bridging midway between the outer wall and girder for both the 10-ft and 14-ft spans is required. In addition, blocking for two non-load-bearing partitions must be included as shown. All other non-load-bearing partitions fall over joist placement.

In plan *B* the girder is centered 12 ft from front and back. This arrangement eliminates the need for using 2 X 10 floor joists whether on-center spacing of 16 or 24 in. is used. However, a problem does exist; the load-bearing partitions rest on the joists approximately 2 ft on either side of the girder. This places horizontal shear forces on the joists. The entire weight of the second floor is distributed on these load-bearing walls and first floor joists and is not borne directly onto the girder.

There is a material savings in dollars if this system is used, but because of the offset of load-bearing walls the placement of the girder is poor. Further, the girder used on the second floor bears on the joists of the first floor rather than on the lower girder.

In plan *C* the girder is off-center toward the front of the building. This arrangement parallels the arrangement in section *A* where load-bearing walls bear directly on the girder. All parameters defined in plan *A* apply here as well. It thus follows that larger joists (2 X 10s) should be used of 24 in. o.c. spacings are employed, the girder may be a 4 X 8 or 4 X 10 depending on how it is used and the sizes of the joists; bridging as well as solid blocking for non-load-bearing partitions must be included. A problem will surface when the second floor is installed.

Second-Floor Assembly Plans

The design criteria for the second floor begins with a 30 psf LL and 10 psf DL. The alignment of joists is directly over the first-floor joists, front to back of building, and at the same on-center spacings. Since there are spans over the dining area, kitchen, and living rooms that have no partition for supporting ceiling joists, girders must be included in the design. Further, a stairwell must be built. The right side of Figure 5-22 shows the second-floor plan and plans A, B, and C.

In plan A two short girders may be built of 2 × 8s and placed on the load-bearing wall. A more desirable arrangement could be a single built-up 4 × 8 girder that spans both rooms. The girder or girders must be made flush with joists as method A in Figure 5-3 shows, so that ceiling drywall and subflooring can be installed evenly on opposite edges of the girder and joists. A review of the Table 1 data in Appendix C shows that No. 2 grade 2 × 8s at 16-in. o.c. spacing should be used, or 2 × 10 grade No. 3 joists should be used if 24-in. o.c. spacing is desired. Notice that the maximum span for 2 × 8s spaced 24-in. o.c. is 13 ft 4 in. and that a Dense Sel Str or No. 1 dense grade is called for. It is quite possible that the span may fall within the limits considering the width of the outer wall (6 in.) and exact replacement of the girder. In the event that the span is not exceeded, the second floor may be framed with 2 × 8s, and where the span is a full 13 ft 4 in. a few special joists need to be used. Elsewhere, No. 2 and No. 3 grade joists may be used.

Plans B and C in Figure 5-22 do not offer any better solution than plan A. But notice also that the load-bearing walls are not over the first floor girder except in the closet area at the right. In plan C a long, 14–15 ft girder is placed over the living-room area. This is a long span for a girder and should be avoided. Since 2 × 8 joists are expected to be used, a steel *I-beam* or larger (6 × 8) wood beam needs to be used.

The second floor joists need to be bridged just as the first floor joists were. Two rows of bridging are installed, one each in the span between outside wall and girder or load-bearing wall. The stairwell needs to be framed in with short headers perpendicular to the joist run and trimmer joists either side of the opening. These are necessary because partitions are installed on each side of the opening.

Conclusion

The plan for the design of the first- and second-floor assemblies of the house in Figure 5-21 points to using plan A as the better alternative even though there is a greater cost factor. Pre-planning of this kind would have included this study, and estimates for cost should have included proper selection of material by grade and size. Setting aside cost for the moment, the designs in plan A and even C provide a more sound, reliable floor assembly than B because the forces on the load-bearing walls are more uniformly aligned. Recall that the second floor bears on the first through the load-bearing walls. If the plan B design were selected, loads would be placed on joists instead of girders. This places horizontal shear stress on all joists, which is not desirable. Finally, the selection of joists includes the *load span* and *on-center spacing*. A proper grade selection can be made only after these three criteria are set. However, if the grades needed for 24-in. o.c. spacing are not available in the selected size, the on-center spacing may be reduced to 16 in. to allow use of available lower-grade joists.

PROBLEMS AND QUESTIONS

1. List the three types of loads bearing on a floor assembly.
2. What is a typical live load per square foot for a living or activity center?
       a. 20 psf   b. 30 psf   c. 40 psf
3. Using Figure 5-1 what is the total pounds per square foot design for upstairs?
       a. 40 psf LL + 10 psf DL   b. 30 psf LL + 10 psf DL
4. When would a girder such as one described in method A (Figure 5-3) be used?
5. List some typical sizes of girders.
6. A building is 24 ft wide × 32 ft long, and a girder is placed midway (12 ft) for the full length of the building and three piers are used. How many feet on each side of the girder is supported by the girder? If the

live and dead load equals 40 psf what would the total weight be on the girder in each 8-ft section?

7. What grade of lumber would you select if a 4 x 6 beam were to bear a load of 40 psf LL and 10 psf DL and span 10 ft?

    a. standard   b. construction   c. No. 3

8. What would be the reason for using a grade No. 2 4 x 8 beam if its loads would be 30 psf LL and 10 psf DL?

9. What is the basic difference between an L-type sill and a box-type sill? Which is used more frequently?

10. Why is mortar used sometimes when a sill is being fastened to a block wall?

11. What causes a squeaky floor assembly?

12. Draw a framed opening in a floor assembly and label the headers, tail joists, and trimmer joists.

13. Why could "cats" be used under floor sheathing and perpendicular to joists?

14. What is the building industry standard for the on-center separated maximum limit for bridging?

    a. 6 ft   b. 8 ft   c. 10 ft   d. 12 ft

15. What grade of plywood would you use as a subfloor if its thickness were 3/4 in. and the joists were spaced 24 in. o.c.?

    a. group 4 CC exterior   b. group 2 CD interior
    c. group 1 Structural II

16. Refer to Figure 5-22. Which condition, A, B, or C allows for the shortest girders upstairs *and* aligns the girders downstairs and upstairs?

17. Explain in a paper why 2 x 10s are used for joists in option A of Figure 5-22.

6

Walls and
Wall Systems

Each wall of a house is designed and built to perform a useful function. It may be one of the exterior walls that defines the shape of the house and includes all the different structural requirements for door passages and window openings, for carrying loads or not, and for having the strength and rigidity required to resist forces placed upon it. It may also be an interior wall whose function is to divide spaces for the purposes of sleeping, living, bathing, and others. The interior wall may need to be designed for load or non-load bearing, for the control of sound, or to accommodate plumbing and heating equipment.

The various types of walls that may be included are those used in single-story construction or two-story construction or may be specialized systems.

These purposes or functions of walls are the subject of this chapter. So that their importance is understood, each wall will be described in terms of a system since many parts are used in special ways to create it. First a definition of a wall system is developed; following this, a look at the elements of the system are defined and discussed; then different wall systems are explained.

DEFINITION OF A WALL SYSTEM

The opening paragraphs defined the purposes and uses of many walls used in residential construction. If the purpose were the only thing necessary to understanding the construction of a wall the carpenter's task would be quite simple. In fact, however, the design of a sound wall system requires considerable knowledge on the part of the carpenter.

Therefore, a more practical definition of a wall system is necessary. For example, a definition that relates the elements of design and construction that make up a proper wall assembly would have much more meaning. Thus, a wall system by definition is one that includes the proper sizes and grades of lumber, includes members such as studs, plates, headers, braces and trimmers, and attaching materials. The definition also includes design data so that when properly constructed the wall forms an assembly that (1) has the proper proportions as dictated by the building plans, (2) has sufficient rigidity to withstand the external forces of wind, earthquakes (where prevalent), (3) provide support for wall coverings, (4) be able to support loads placed upon it, and (5) incorporate any and all special features that may be required.

WALL MATERIALS AND THEIR USES

Studs

Throughout the history of house construction the stud has been cut in many sizes and shapes. In early times in this country and in Europe, studs approximately four inches square were used in the following places: (See Figure 6-1.)

1. at each corner
2. each side of a window and door
3. between long expanses
4. for corner braces

Figure 6-1 Old-fashioned framing with 4 x 4s.

Then mortar, clay, straw, bricks, stone, or any material available was plastered between the studs to complete the wall.

Only recently has the stud become uniformly sized and even more recently its size has been reduced slightly to its present 1-1/2 in. x 3-1/2 in. (approximately 40 mm X 90 mm in metric) as Figure 6-2 shows. Two common lengths are used in wall systems and they are the approximate 8-ft. length for single-story, platform, or western framing techniques, and the 16–18 ft stud for balloon framing.

Figure 6-2   Size of a modern 2 x 4 stud.

Present construction requirement of placing studs at either 16-in. or 24-in. o.c. spacings has several desirable features. First, the quality of stud or its grade may be selected from poorer stock since the load it must carry is distributed to many studs. Consider the condition in Figure 6-3 where a stud is in a *load-bearing wall.* This view shows that a single stud would be a poor selection as a support for such a load. This situation is frequently seen where a porch roof is temporarily supported by studs while columns are being readied. However, even the best quality stud would show some form of weakness in this situation.

In the figure the stud is shown under a heavy load that compresses all of its fibers downward. Since the grain of the stud may not be uniformly straight and there may be knots and places of tension and compression within the stud, it will *bow.* It is probable that even the best grade stud made from the strongest species—Douglas fir larch or southern pine—would bow under such a load.

Figure 6-3   Total load bearing on a single stud and the effects on the stud.

A common stud wall will not have one stud supporting such a load. As Figure 6-4 shows, where studs are evenly separated and are placed directly under the force of the roof or joists the compression is evenly distributed among the studs. Therefore, the studs are more capable of retaining their original shape and will not, as a rule, bow.

However, there is a possibility that some deflection may take place and the grade of lumber that is used will allow or limit this deflection. Figure 6-5 shows four grades of stud and their relative usefulness in supporting loads. The better the grade of stud the more stress it can stand; the poorer the stud the less stress it should have placed on it.

In usual framing practices the *utility* stud suffices for both load-bearing and non-load-bearing walls even though its grade is poor. The reasons for this are the close on-center spacings of the studs, wall bracing, double plates, and relatively small ceiling joists and roof rafters. If larger ceiling joists or roof rafters and heavier live loads are predicted, better grades of studs are used to ensure that the heavier loads will be supported properly.

Distributive force in multi-stud wall

Figure 6-4 Total load on a stud wall with equal distribution of force per stud.

Figure 6-5 Relative ability of various grades of studs to withstand force of loads.

Seven grades of studs are listed in Table 6-1. Of the seven the first one, No. 1 and appearance, and the second, No. 2 may be used where exposed studs are needed. The other grades usually have objectionable qualities such as knots, stains, pitch pockets, streaks, shakes, splits, and other defects that prohibit their use as finishing materials. The other grades do, however, meet the building requirements of most local, state, and national building codes. There is a restriction on the use of the *utility* stud; its use is limited to 16-in. o.c. spacing in wall construction. This restriction corresponds to the earlier explanation illustrated in Figure 6-5.

All studs must be dried properly so that their moisture content is either 19 percent or, if kiln dried, 15 percent. This is important even though shrinkage in the length of a stud is the smallest shrinkage that takes place. If quantities of moisture exceed these values, as they would in green lumber, *settling* will result as the studs in the wall dry out. This would cause floors and other level surfaces to become uneven. A one-tenth of one percent shrinkage in an 8-ft stud causes less than 1/4 in. reduction in length; but in balloon framing, where 18-ft studs are used, the shrinkage of green lumber could exceed 1/2 in. Therefore, use only cured studs in wall construction and eliminate the problems green lumber could create.

TABLE 6-1   Studs by Grade and Use

| Grade | Type of Wall | | Range of | |
	Load Bearing	Non-load Bearing	Stiffness E (10)	Strength F_b (psi)
No. 1 and appearance	Yes[a]	Yes[c]	0.9–1.8	1050–2050
No. 2	Yes[a]	Yes[a]	0.9–1.7	1000–1650
Construction	Yes	Yes	0.8–1.5	725–1200
No. 3	Yes	Yes	0.8–1.5	475– 925
Stud	Yes	Yes	0.8–1.5	550– 925
Standard	Yes	Yes	0.8–1.5	350– 675
Utility	Yes[b]	Yes	0.8–1.5	175– 325

[a]May be used where exposed studs are needed.
[b]May be used when o.c. spacing is not greater than 16 in.
[c]May be used but often exceeds needs for strength and stiffness.

Since studs are nailed to plates as shown in Figure 6-6 they need to be nailed securely. Two methods are used; one uses two 16d common nails driven through the plates into each end of the stud. The other method uses four 8d common nails toenailed into the stud and then into the plate. When toenailing, 60 degrees is the best angle to use as it allows the nail to adequately penetrate the stud, yet permits most of the nail to anchor into the plate.

Figure 6-6   Nailing of stud to plate either through plate or by toenailing.

Plates

The plate or horizontal member in the wall assembly has almost always been used on top of the studs as a way of holding studs in line. The plate has not always been used at the base of the studs. However, the plates are a very important part of the wall assembly in that (1) they do perform the need of keeping the studs in alignment, (2) they provide nailing surface for wall covering materials on both sides, and (3) they perform the function of *fire stops*.

Throughout the history of construction, the plates have been called by various names. Some of these are the *sole, sole plate, shoe, shoe plate, sill plate, bonding plate, connecting plate*, and *top plate*. All but the bonding, connecting, and top plates are names that refer to the plate used at the base of a wall assembly. More currently the term *plate* refers to either the bottom or top horizontal members of a wall assembly.

Figure 6-7 shows the position of the plates and also shows that the plate performs the three functions cited above. The current trend is to use a double or built-up plate at the top of the wall. This is done not so much for the strength needed to hold the studs in alignment, because a single plate can do this, but because better overall building lines are established and the top plate (bonding plate) is used to unite other wall assemblies through overlap construction (Figure 6-7). In addition, the second top plate also aids in supporting horizontal shear stresses that might occur, and it is usually recommended for load-bearing walls.

Figure 6-7 Three functions of plates (1) maintain stud separation, (2) nailing surface for paneling, and (3) fire stops.

These reasons are sound design principles put into practice; yet several studies have shown that the single top plate has been used in non-load-bearing exterior and interior walls with no deficiency in wall quality.

Plates usually are nominal 2 × 4s and should be of a grade equal to or better than the stud grade. They should not be a poorer quality lumber. Plates must be properly seasoned so that their moisture content is either 15 percent (KD) or 19 percent. The reason for this is to minimize the problem of shrinking. The shrinking shown in Figure 6-8 could create problems with paneling and siding because of the tension placed on these items. Panels cannot shrink; they can only buckle.

Figure 6-8 Shrinking as a problem from using improperly cured lumber.

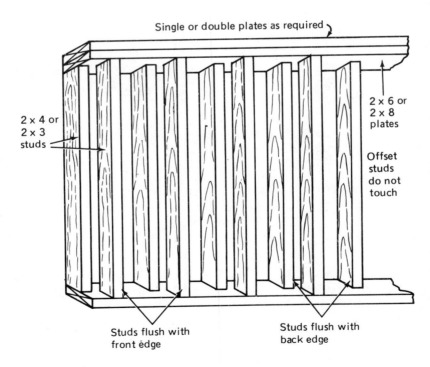

Figure 6-9 Double wall single plate—2 x 6 or 2 x 8.

Double wall plates are frequently needed when a soundproof wall is desired. These plates are usually 2 × 6s and, as Figure 6-9 shows, either 2 × 3 or 2 × 4 studs are off-centered so that their surfaces do not contact. Because of their width, the plates allow each stud standing room so that sounds received by one stud are not passed to the other and into the adjoining room.

The 2 × 6 plate may also be useful in wall construction where plumbing vent pipes and sewer pipes are installed above floor level. They will hold 2 × 3, 2 × 4, or 2 × 6 studs in alignment, and may be cut or bored for passage of the plumbing pipes and still retain considerable strength.

Headers

Window and door framing requirements in load-bearing exterior walls and some non-load-bearing exterior walls require the use of a header. Load-bearing interior partitions also need headers where passageways from room to room are framed.

The header, the horizontal member spanning a desired open space, is customarily made from nominal 2-in. stock and installed on edge as Figure 6-10 shows. The various sizes employed to support the forces and weights resting on them are shown in Table 6-2. The table provides conservative maximum spans for headers. Local building codes may vary somewhat either way.

Figure 6-10 Headers—dimension and length.

TABLE 6-2 Spans for Headers

Header Size	Maximum Span
2 2 x 4s	4 ft. 0 in.
2 2 x 6s	5 ft. 6 in.
2 2 x 8s	7 ft. 6 in.
2 2 x 10s	9 ft. 0 in.
2 2 x 12s	11 ft. 0 in.

Note: All members are placed on edge.

The importance of a header is illustrated in Figure 6-11. The roofing materials and live roof loads, ceiling joists, ceiling materials and live ceiling or attic loads bear on the wall. This means that the loads are directed downward to the headers. Notice carefully that the concentration of the load is at each end of the header where the force is then transferred to the floor assembly. This conventional approach illustrates that the window or door unit does not bear any of the weight of the wall, and this is as it should be. It also illustrates the importance of using the correct size of header.

Figure 6-11 Forces on headers and distribution of forces to floor.

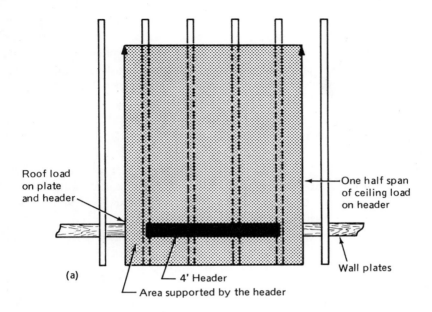

Roof load
on plate
and header

One half span
of ceiling load
on header

Wall plates

(a)

4' Header

Area supported by the header

TOP VIEW

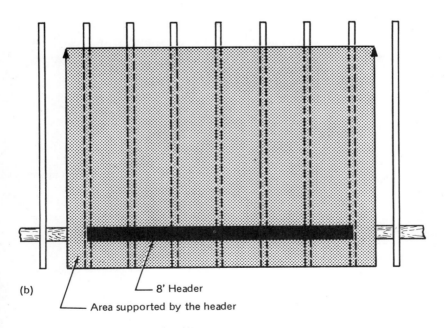

(b)

8' Header

Area supported by the header

Figure 6-12 Load bearing on headers of different lengths.

As the span of header increases, the load placed upon it increases. It has already been shown in the study on girders (Chapter 5) that the total loads multiplied rapidly as the spans increased. The same holds true for the header. Figure 6-12 provides two sample situations that illustrate this fact. In Figure 6-12a the blackened area is only 4 ft wide and represents the header. The shaded area is the load area or total load the header must support. Notice that it is one joist space wider than 4 ft and equal to half the joist length plus the rafter force bearing upon the wall and header. In Figure 6-12b the span is increased to 8 ft. Notice how much more total weight the header must carry. Headers made from doubled 2 × 10s must be used to span the 8 ft as Table 6-2 indicates.

A recent adaptation or modification to the placement of headers is shown in Figure 6-13. Instead of the header being placed at door or window rough opening height, it is placed snugly under the wall plates. The same principles of load bearing are still included in this arrangement. Instead of cutting cripple studs to fill in between headers and plates, a single 2 × 4 is installed across the top flat-wise at rough window or door height to provide nailing surface for wallboard, sheathing, window head nailing surfaces, and window or door trim. This technique may not look as sound as the conventional one but it is, and it may save labor and material costs.

Figure 6-13   Relocation of header snugly under top plate.

The preceding discussion on headers has indicated the important function they serve in supporting loads, and yet there is one more property of a nominal 2-in. piece of lumber that should be taken advantage of when constructing headers. The *crown* of each header piece should be placed toward the top plates (similar to crown-up on floor joists). As the weight of the load is applied to the header, it will deflect downward. The crown-up method of construction is effective in reducing the deflection caused by the bearing load.

Finally, properly dried lumber of good quality, grade, and species should be used for headers. If this type of lumber is used shrinkage is minimal. However, if green or partially dried lumber is used for headers two conditions result: First, the members will dry within the wall and significant shrinkage occurs. This has a severe effect on rigid panels fastened to the wall or brick veneer on the outside of the wall because they do not shrink. Buckling or other structural rupture may result.

The second reason to avoid using unseasoned lumber is that there is a tendency for the fibers in the *bearing zone* to collapse, become crushed, and cause settling. Even before drying out takes place the damage is done. So to avoid the problem of shrinkage for either reason use only properly dried, good quality lumber for headers.

Trimmer Studs

The trimmer stud is one that extends from floor (sole plate) to ceiling (top plate) on either side of the openings in the wall. Figure 6-14 shows examples of these studs. A *trimmer stud* performs a function very much like the trimmer joist in a floor assembly around an opening. As a rule, the headers are nailed through the trimmer. The trimmer bears a load placed upon it from above in a load-bearing wall, but does not assume a significant part of the load of the header.

Trimmers are used in non-load-bearing walls in similar fashion. There too the headers are nailed through them; and, in either type of wall, the trimmer may be a normally spaced stud or an additional stud used exclusively for this purpose.

Figure 6-14   Trimmer studs.

Cripple Studs

A *cripple stud* is one that is used in several different places in a wall assembly. It is customary to use cripples in vertical positions. They may be spaced at normal on-center separations as in a cripple wall, or above headers as in Figure 6-15. In both of these applications they act as short studs. They may be held in place by either spiking through plates or toenailing as any other stud is fastened.

The third purpose for a cripple stud is its use in framing the opening of a door or window as Figure 6-15 shows. The cripple stud (sometimes called a *jack* stud) is fit beneath the header. This means that it is the bearing surface for the header and is also the member that bears one-half the total header load to the lower plate. It is, therefore, very important that this member be made of good quality lumber and be properly dried. Even though it is nailed securely to the trimmer it would shrink if unseasoned, thereby causing stress within the wall and on materials attached to the wall.

Figure 6-15 Cripple studs.

In window framing, the 2 X 4 installed as a sill framing member should be toenailed to the cripple. Many times carpenters use short cripples and extend the sill to the trimmer. This arrangement has the effect of making 1-1/2 in. on each end of the sill a part of the load-bearing area from header to floor.

Braces

Almost any material installed within or onto a wall framed of 2 X 4s or larger stock, braces the wall. Bracing is needed to control and prevent a building wall from *racking*. If a wall is out of plumb measured along one end from sill to plate it is racked. If such a condition is found, racking was allowed to happen because of no bracing or inadequate bracing.

Many times a wall that does not contain diagonal bracing or let-in bracing is mistakenly considered unbraced. Actually, though, ordinary 1 × 6 or 1 × 8 sheathing properly installed provides bracing. If sheathing is installed horizontally, as in Figure 6-16a, it has some ability to resist the racking forces of wind and earthquake. However, its function as a brace is limited and custom has dictated that specific bracing such as the let-in brace of a 1 × 4 be included, as shown in Figure 6-16b. On the other hand, when sheathing is installed diagonally at about a 45-degree angle as Figure 6-16c shows, better bracing is obtained. Sheathing properly fit and well nailed with three 8d or 10d common nails per stud crossing does indeed resist the forces that cause racking.

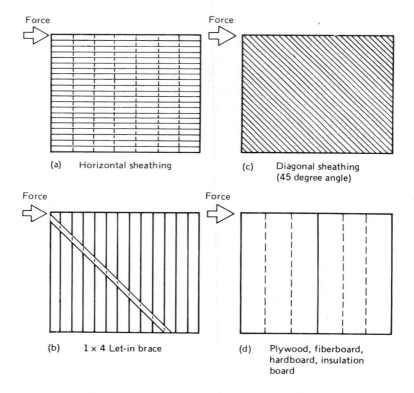

(a)   Horizontal sheathing

(c)   Diagonal sheathing (45 degree angle)

(b)   1 x 4 Let-in brace

(d)   Plywood, fiberboard, hardboard, insulation board

Figure 6-16 Bracing the wall to resist forces.

The use of diagonally installed sheathing boards has largely given way to panels made of many materials (Figure 6-16d). Cost of labor is one reason for the switch, since many hours are needed to cut and install diagonal sheathing, but a panel may be installed in minutes. Cost of materials is a second reason, but more importantly the panels (4 ft × 8 in. or other) properly installed improve bracing up to eight fold.

The first trend in using plywood panels as bracing was to sheath the entire wall with plywood. Sometimes the panels were installed horizontally, other times vertically. Given one way or the other, tests have proven that the vertically installed panels provide more bracing. Next came the use of plywood panels at all interior and exterior corners around the building. Fiberboard panels were installed between plywood panels; but although these had some insulating ability they had very little racking resistance.

Manmade panels (other than plywood) have been improved through better combinations of materials and improved manufacturing techniques. Some of these panels have resistance qualities better than horizontal sheathing boards. These are used entirely around a house, not just between corner panels of plywood.

Still, however, the best bracing of a wall frame is obtained by using plywood, glue, and nails. Panels preferably full length and full width are glued and nailed to the studs and plates. Table 6-3 illustrates in descending order the bracing quality of each type of bracing material.

Several obvious factors are indicated in the table, and several are not. Panel type and material are as important as panel thickness. Plywood, for example, may be made from group 1, 2, 3, or 4 species, and each panel made from a different group would have a different relative racking resistance. The group 1 panels have the greatest strength and stiffness, so they would provide the greatest resistance to racking forces. Group 2 panels provide less resistance, as do group 3 panels, and group 4 panels provide the least racking resistance.

Tempered hardboard, which is a manmade product, has remarkable racking resistance. Fiberboard, which is also manmade, is only one-third as good as hardboard, but still three times better than horizontal sheathing. Then too, some insulating panels have some rack resistance capability, but nowhere as much as plywood or tempered hardboard.

TABLE 6-3   Braces for Framed Wall

Bracing Material	Relative Racking Resistance[a]	Relative Strength	Notes
Plywood (glued)			Installed vertically
5/8 in.	24	9+	
1/4 in.	14	8+	
Tempered hardboard, 1/4 in.	9.8	6.9	Nailed 3-in. and 6-in. spacing
Diagonal sheathing			45-degree angle
1 x 6 or 1 x 8	7.3	8+	Studs 16 in. o.c.
Plywood (nailed only) 5/8 in.	4.8	9+	installed vertically
1/4 in.	4.2	5.2	16-in. o.c. or 24-in. o.c. studs
1 x 4 let-in brace			45-degree angle brace.
	4.4	3.9	16-in. o.c. studs
	2.7	2.0	24-in. o.c. studs
25/32 in. fiberboard	3.0	3.8	3-in. and 6-in. nail spacing
	4.5	6.1	Double nailing
Metal straps	1.9	1.4	2 each, nailed 45 degrees at corners with two nails per stud
Horizontal sheathing 1 x 6 or 1 x 8	1.0	1.0	Values may be increased up to 50% by increasing nail size to 10d and using more nails

Source: Data extracted from "Guide to Improved Framed Walls for Houses," U.S. Dept. of Agriculture, Forest Service, 1965.

[a]Values are stated for solid wall with no window or door cutout in first 4 ft from corner.

Nailing and gluing also play an important role as has been previously mentioned. Table 6-3 states the nailing schedule for perimeter first and intermediate studs second (note column). Finally, siding may contribute much to the rigidity of a wall if it is a panel type; if it is a lap siding variety, the siding does not.

Sheathing and Wall Coverings

The types of sheathing identified in Table 6-3 were plywood, fiberboard, hardboard, and 1 X 6, or 1 X 8. Insulating panels made 1/2-in. to 2-1/4-in. thick are being used directly over studs. Their function is insulation which, of course, is vitally important today. As seen in Figure 6-17, let-in corner bracing is used, and the sheathing-insulating board is nailed directly to the studs.

Use of thermal boards as sheathing makes for a wall system that may exceed 6 in. in thickness. Chapter 3 explained the importance of these new products in thermal control. But, when siding is installed over thick insulating panels very long nails are needed to secure the siding to the stud. The insulating panel does not have nail-holding power.

Figure 6-17 Sheathing a wall with insulating sheathing (*Courtesy of Celetox Corp.*).

On the inside of the wall a panel of some sort is usually installed. It may be 1/2-in. gypsum wallboard, 3/8-in. gypsum wallboard and panel boards, lath and plaster, or sound resistant panels and gypsum wallboard. Each of these products adds:

1. To the overall dead load of the wall,
2. Strength to the wall,
3. Improvement in its racking resistance, and
4. To its thermal protection and sound control.

In every case proper nailing and adhesives improve the total quality of the wall.

Each and every product or part of the wall previously described contributes to the development of a soundly constructed wall system.

EXTERIOR WALL SYSTEMS

All, most, or some of the uses of materials described in the preceding section may be made part of an exterior wall system. If the wall is load bearing, most of the materials will be used and their sizes, grades, nailing schedules, and other factors must be properly incorporated. If the wall is non-load bearing, not all materials need be used, primarily because bearing loads are not present.

Within the descriptions and illustrations that are to follow, design characteristics and proper application of materials are presented. Segments of walls are presented first then a full wall is shown from the floor plan and elevation plan.

Balloon Framing

The studs in a balloon frame extend from the sill to the double plates that the rafters and second floor ceiling joists rest upon. Figure 6-18 shows a corner segment of a balloon frame. Many of the design features identified earlier are incorporated here. They are:

1. *Let-in corner braces.* Let-in braces of 1 × 4 extend outward and downward from the corner in each wall. Before the wall is sheathed a second pair may be installed in the

Figure 6-18 Balloon frame wall construction.

second floor segment of stud or may be omitted if plywood or other properly rated panels of sheathing are installed.

2. *Fire stops.* Two sets of fire stops are installed; one is above the first-floor joists to trap any basement fires. The other is installed above second-floor joists to trap first-floor fires.

3. *Studs rise two full stories.* Each normally spaced stud and all trimmer studs rise full height. They need to be one of the seven grades listed in Table 6-1. Certain building codes may restrict or prohibit use of the utility stud.

4. *Plates are used at each end of the stud.* A single sill plate is anchored to the foundation. A double plate is constructed around the top of the walls. Overlapping at corners and partitions is accomplished with the bonding plate.

5. *Ribband (ribbon-band).* A 1 × 4 ribbon band is let in horizontally along the wall perpendicular to the second-floor joist run. Second-floor joists (beams) are notched to fit over the ribband to lock wall to joist (see inset in Figure 6-18). Nailing is important, but the notch prevents shifting and separation. In other words, an interlock union is made between wall and floor assembly.

6. *Second floor ceiling and roof rafters.* Both the joist and rafters bear on the plates, which are aligned over studs. Good design for load and force is shown.

7. *Common inside-outside corner.* A commonly designed corner is constructed from three 2 × 4s with scrap pieces used as fillers.

The balloon framing method is an excellent one to use for two-story construction. It is also very appropriate for split-level house construction. It is also desirable to substitute a balloon frame for a cripple wall supporting a platform wall.

Platform, Single-Story, or Western Framing

The most commonly used framing technique or method is the *platform*, sometimes called single story or western. Its wall assemblies are one story high; this usually means an 8-ft ceiling height. Its popularity is founded on its simplicity of construction

Figure 6-19 Platform, single-story, or western framing.

and ease of erection. Many of the design features defined earlier are incorporated in the sample of platform framing shown in Figure 6-19. They are:

1. *Plates*. Sole and double-top plates are used on all perimeter walls. The top plate is held in alignment by the bonding plate.
2. *Corner braces*. Let-in 1 X 4 corner bracing is used on each wall. (*Note:* panels may be substituted for let-in braces.)

3. *Trimmer stud*. Trimmer studs are used either side of window or door rough openings.
4. *Headers*. Properly sized headers are installed edge up, and supported on each end by cripple studs that extend downward to sole plate.
5. *Cripple studs* are used around window opening, above header, below window sill, and supporting header.
6. *Common inside-outside corner*. A common three-stud corner is constructed with scrap 2 × 4 spacers between studs.
7. *Ceiling joists* align over studs.

Blueprint Data and the Wall

Blueprints are the fundamental link between the architect and carpenter-builder. The set of plans relates the details necessary for the building's shape, size, and other factors. With a set of plans the carpenter constructs the framework of the building. Two of the most useful pages in the set of plans are the floor plan and elevation and detail drawings. However, the plans are not the only link between architect and carpenter or builder. Both know that building codes specify certain requirements and both know that the carpenter must fulfill the requirement of constructing a sound wall system.

To show the relationship of the plan to the final wall system, the next several paragraphs and illustrations examine a platform construction technique of an exterior load-bearing wall. First, the wall is shown with conventional 16-in. o.c. studs. Then the same wall is shown with 24-in. o.c. studs with modular spacing.

The floor plan segment shown in Figure 6-20 provides necessary dimensions for (1) wall length, (2) position of the windows and door, (3) location of offset in wall and (4) points where interior partitions intersect with exterior wall. Usually a scale of inches per foot is provided so that the carpenter can locate the exact position of each window and door. The sizes of windows and doors may either be listed on the floor plan (not usually), written into the elevation plan (sometimes), or provided in a list or schedule attached to one of the plans.

The elevation plan also shown in Figure 6-20 shows (1) the window shape and height, (2) door and adjoining window

Figure 6-20 Blueprint data and framed wall—conventional 16-in. o.c. stud separation.

Bonding or connecting plate

Ceiling joist

Rafter

Let in corner

Corner stud assembly

2 x 4 sill

Header

Conventional 16" o.c. load bearing exterior wall

Elevation plan (front wall)

BRICK

Floor plan (front wall)

5"

16' 0"

24' 8"

70' 8"

71' 6"

14' 0"

16' 0"

5"

137

sidelights, (3) roof shape, (4) cornice line and design and exterior wall material (brick). Notice that the windows and door heights are aligned under the overhang of the roof. This translates to a lower header edge of 6 ft 10-1/2 in. from the subfloor, since this is the dimension necessary for door unit installation.

The carpenter precuts a quantity of studs based on three per 4 ft of wall for the conventional 16-in. o.c. stud spacing. In addition, extra studs are cut for trimmers and corners. Recall that trimmers are full-length studs placed either side of door or window openings. Corners are made in various ways, as Figures 6-21 a, b, and c show.

When using the conventional 16-in. o.c. method of wall construction (Fig. 6-20) no special consideration is given to the placement of windows or doors other than to follow the plan. It therefore may be that the trimmer stud is exactly on line with 16 in. o.c. or it may be off-centered any distance. Many additional trimmers, and cripple studs as well, may be needed to complete the wall.

By requirement, the headers for all three windows will be constructed from 2 × 4s set edge up. However, the door opening is 8 ft wide and needs to have a header made from 2 × 10s. The builder may elect to use 2 × 10s for all headers for uniformity of construction even though they may exceed requirements over windows and single door openings. Each header must be end supported by 2 × 4 cripples, and cripples 16 in. o.c. need to be included between header and top plate.

Let-in corner braces are used. The one on the left is at a 45-degree angle. The one on the right is installed at an angle more than 45 degrees since a window would interfere with it. If the one on the right were installed at a 45-degree angle its effectivness would be lost.

The partitions intersecting with the wall are bonded to the front wall by the *bonding* or *connecting* plate. Interior corners also need to be constructed as Figures 6-21d and e show.

The ceiling joists are aligned directly over studs or not more than a 2-in. thickness from the stud beneath. The rafters are butted to the joists, and they are either resting directly over the stud or just 2 in. from the stud beneath it.

The same floor plan (or one almost like it) can be used with the modular, 24-in. system. This system is slightly more involved

Figure 6-21 Types of corner assemblies for exterior and interior corners.

139

Corner bracing

Single plate non-load-bearing wall

Header

4' 4' 4' 4' 4' 4' 4' 4' 4' 8' 4' 4' 4' 4' 4' 2'8"

4' Modular units
24" system

Notes:

Window and door openings
fall within modular units
2' 8" should be reduced to 24"

Headers for windows - 2 x 4

Header for door - 2 x 10

Figure 6-22 Framing a wall using the 24-in. modular system.

than just spacing the studs on 24-in. centers. It involves a design principle of modules. A module may be a minimum of 4 ft wide, or wider, such as 6, 8, 12, or 16 ft wide. Usually modules are designed as 4, 6, and 8-ft widths. Figure 6-22 illustrates the exterior wall of the floor plan in Figure 6-20.

Several changes to the floor plan are needed to make it conform to the 24-in. system. First, the window placement needs to be shifted so that the windows fall within a 4-ft module. The door unit also needs to be shifted to fall within an 8-ft module. Figure 6-22 does not show this idea very well, but Figure 6-23 does. Notice how one trimmer is always at the normal 24-in. o.c. position in the 24-in. system.

Figure 6-23 Framing window off module and on module.

Another change needed to be made is the last egment of the wall (right side). It is a fragment of a module (2 ft 8 in.). This should be changed to either a 6-ft module with the window as part of the module or another full 4-ft module. The plans would need redrawing. Notice that plywood or other suitable panel sheathing is used for corner bracing. If the right end of the wall were expanded to a 4-ft module, a full sheet of sheathing would be effective as corner bracing there also.

A double top plate is used in the 24-in. system for alignment and connecting partitions. A single plate may be used for partitions that are non-load bearing. Joists and rafters are spaced 24 in. o.c. and they need to be located in line with studs as described earlier in conventional framing. A reminder needs to be made here; with joists and rafters spaced 24 in. o.c., larger members must be used. Be sure to select sizes appropriate for loads. The studs identified in Table 6-1 should carry the loads of these larger members.

INTERIOR PARTITIONS

Interior walls are classified as either load bearing or non-load bearing; they are also referred to as *partitions*. A partition therefore may be considered to be an assembly that is designed to serve a particular purpose. If the partition is a load-bearing type, its construction must follow the same design principles as the exterior load-bearing wall design. Headers of proper sizes must be used. Studs must be centered so that ceiling joists bear directly over them when nailed to the top plate. Double plates need to be used. Door or other passage openings must have trimmer studs and cripples installed as parts of the assembly. As a rule no corner bracing is included in the assembly.

If the partition is non-load bearing its construction is simpler. It may be built with single plate at bottom and at top. If this is done studs must be cut 1-1/2 in. longer to allow for the one top plate vs. two top plates. Headers do not support anything, so they may be made from small materials such as 2 X 4s or 2 X 6s. Cripples should be used along side door openings to provide a sound nailing surface for door jambs and door trim (see Figure 6-24). Cripple studs above headers may be omitted if dry wall is installed horizontally. On the other hand, if paneling is installed vertically and a joint is made over a header a cripple stud must be installed (Figure 6-24b).

Figure 6-24   Interior partition, non-load-bearing construction details.

Interior walls usually are constructed from 2 × 4s; however, they may be built using either 2 × 2s or 2 × 3s and 2 × 6s or 2 × 8s. The 2-in. or 3-in. wall is found in and around closet and chimney areas. They are not as sound as the normal 4-in. wall, but where little chance of external force is expected they may prove to be space saving. The 2 × 6 or the 2 × 8 wall is usually built where soundproofing is needed or where plumbing installations must be accomodated. Figure 6-25 shows a wall with plumbing pipes and helps explain the need for the wider plates and studs.

All of the factors mentioned in the preceding paragraphs are equally important in either the conventional 16-in. o.c. stud wall system or in the 24-in. modular system. Modules are used for partitions just as they are used for exterior walls.

Figure 6-25   6-in. or 8-in. partition housing plumbing vent and water pipes.

Summary

A wall system is an integration of many products, structural members, and sound design principles. As this chapter points out, no one item by itself can adequately support a load or perform a function. Collectively, however, if properly selected, installed, and secured, the products and members are united into a sound design that does all that the requirements call for.

More emphasis has been placed on the fundamental structure of the wall, its studs, header, plates, and bracing than on the products attached to it because the structure is the skeleton and it must be designed correctly. The sheathing, paneling, sidings or interior drywall, lath and plaster, and paneling all contribute to the wall's ability to perform its function. Recall that these may be resistance to wind, earthquake, sound, heat or cold. Each problem that must be reckoned with adds to the wall's quality.

Finally, a wall may be constructed conventionally with studs spaced 16 in. o.c., windows and doors located in the most desirable places, and intersecting partitions joining the outer walls wherever desired. Considerably more materials are needed to construct this type of wall than one designed in the 24-in. modular system. The modular concept of wall building reduces the quantity of materials used for its construction, thus saving time and money. It is as sound as a conventional wall design because of the method of construction.

Although the conventionally designed wall will be used for many years, wall construction is moving toward the modular 24-in. system.

PROBLEMS AND QUESTIONS

1. What should the exterior wall do for a house?
2. List four purposes for an interior wall.
3. What are the five requirements a wall must satisfy?
4. What are the two most often used on-center spacings of studs?
    a. 12-in. and 16-in.
    b. 16-in. and 24-in.
    c. 16-in. and 32-in.
    d. 24-in. and 48-in.
5. If you were using the metric system, what size would a stud be?
6. What makes a single stud bend if a section of roof is bearing upon it?
7. Why don't studs in a wall bend when loads such as a joist and roof assembly are placed upon them?
8. Which grade of stud would deflect the least under a load?
    a. Sel. Str.   b. No. 2   c. No. 3   d. Utility
9. Draw a section of wall and label the stud, sole plate, top plate, and fire stop.
10. Can utility studs be used as primary studs in a wall with 24-in. o.c. spacing?

11. What two purposes does the bonding plate perform?
12. Load-bearing headers must be made of _____ if the span is 62 in.
       a. two 2 × 4s   b. two 2 × 6s   c. two 2 × 8s
13. Is there any loss in utility if a header is placed snugly under the wall top plate? Do the jack studs bear any load or is the load passed to the trimmer studs? Explain?
14. What is the use of a trimmer stud in framing an opening for door or window:
       a. Carries the load
       b. Reinforces and helps align the jack stud
15. A cripple stud:
       a. Extends from floor to ceiling
       b. Is a short length of stud at stud o.c. spacing
       c. Always used over headers
       d. Is a full-length wall stud
16. Which bracing of a wall is more effective?
       a. 1 × 6 sheathing
       b. Let-in corner boards
17. Which provides the greatest protection against racking forces?
       a. Plywood glued to studs
       b. Diagonal sheathing
       c. 1 × 4 let-in braces
18. Name four qualities 1/2-in. drywall adds to a wall.
19. Describe the main distinctive feature of balloon framing. Are fire stops used?
20. What are two other names for western framing?
21. Using the plan in Figure 6-20, how many common-length studs would you cut for framing the front wall if they are spaced 16 in. o.c.? How many for exterior and interior corners? About:
       Studs   a. 47   b. 54   c. 60
       Corner studs   a. 9   b. 12   c. 16
22. Where would headers be used? (Figure 6-20)
23. Where would cripple studs be used? (Figure 6-20)
24. What alteration would you make to the plan in Figure 6-20 to make it fit a model Mod 24 plan?
25. How many trimmer studs would you save if you placed the three windows and door (Figure 6-20) on module rather than off module?
       a. none   b. 2   c. 4   d. 6

7

The Roof, Part I:
The Common Rafter

This chapter and the next three chapters discuss the various rafters used in roof construction. The common rafter is defined first so that many of the properties of all rafters may be identified and a sound understanding of the function of a rafter is established.

There are many aspects about a common rafter that are important to know so that they may be used by the carpenter. He needs to know where to use a common rafter in a roof assembly versus using a hip or jack rafter. Terms such as pitch, run, rise, and others must be thoroughly understood, as well as the forces the rafters must sustain and the forces placed on walls. Then, too, he must be able to calculate the rafter length after defining its rise and run from plans. Having obtained all this information, the carpenter then must select the proper rafter grade, size, species, and finally determine how to locate it properly on the wall.

DEFINING A COMMON RAFTER'S LOCATION
IN A ROOF ASSEMBLY

Since a roof frame may be made from a variety of different types of rafters, it is important to recognize where common rafters are used. Figures 7-1 through 7-5 illustrate several of the most common roof styles. The shaded area identifies the area in the roof where the common rafter is used.

The roof in Figure 7-1 is the common gable roof design. It is constructed entirely with common rafters as the shaded area indicates. The roof in Figure 7-2 is the *shed* or *low-level pitch* roof design. Each rafter is a common rafter as the shaded area shows.

The rafters in this style roof also act as ceiling joists, since ceiling materials are attached to them.

Figure 7-1   Gable roof made entirely with common rafters.

Figure 7-2   Shed or low-level pitch roof, made entirely with common rafters.

The gable roof shown in Figure 7-3 is constructed with common rafters, as are the dormer roofs. The front dormers use common rafters in the shaded area. The area left white is not constructed from common rafters but uses a type called jack rafters.

The *hip-and-valley* roof shown in Figure 7-4 does make use of the common rafter where the shaded areas show two different locations. They are the areas where rafters are not connected to either hip or valley rafters. This means that the common rafter attaches at the roof peak and rests on the wall plate.

The roof in Figure 7-5 is the *gambrel* roof design. It is constructed with common rafters. One set of common rafters is cut to fit the very steep pitch of the lower section; the other set of common rafters is cut to fit the low pitch of the upper section. The dormer's roof is made with common rafters in shed fashion.

Figure 7-3   Dormer roofs, using common rafters for part of their make-up.

Figure 7-4   Hip and valley roof, using common rafters outside the area of hip or valley.

Figure 7-5   Gambrel roof design using common rafters.

There are other varieties of styles of roofs that are really combinations of one or more of the designs shown in the five figures. They too may use one or more common rafters in their frame.

A clear definition of a common rafter should now be apparent. It is the one that attaches to the exterior wall and reaches to the ridge or peak of the roof. This definition even holds true for the dormer common rafter.

PITCHES AND HOW THEY ARE DEFINED

The use of a specific pitch stated in fractional value has been largely replaced with the more general terms a *low-pitched* roof, a *high-pitched* roof, or a *steeply* pitched roof design. However, the term pitch is understood by all who refer to pitch in any terms. Therefore, a fundamental understanding must be the common ground. All fractional notations (1/4, 1/3, 1/2, etc.) of pitches of roof design relate to its flatness or steepness in terms of percentage or fractional value of total span versus total rise.

The pitch may be a fractional value of a full pitch as shown in Figure 7-6. When related to a framing square the pitch is obtained by doubling the 12 in. (shown in Figure 7-6a), then dividing the 24 in. into the 4 in. for example:

$$1/6 \text{ pitch} = \frac{4 \text{ in.}}{24 \text{ in.}} = \frac{1}{6}$$

$$1/2 \text{ pitch} = \text{double 12-in. run} = \text{24-in. span}$$

$$\frac{12\text{-in. rise}}{24\text{-in. span}} = 1/2$$

The pitch table (Figure 7-6b) is standard; but it too is not easily understood. The table indicates, for example, that a 1/2 pitch has a run and rise of 12. The pictorial just below the table translates the table data. Notice that the rise and run are indeed 12 ft in this case. However, the span is twice the run or 24 ft, and when this value is used the pitch can be calculated.

$$\frac{12\text{-ft. rise}}{24\text{-ft. span}} = 1/2 \text{ pitch}$$

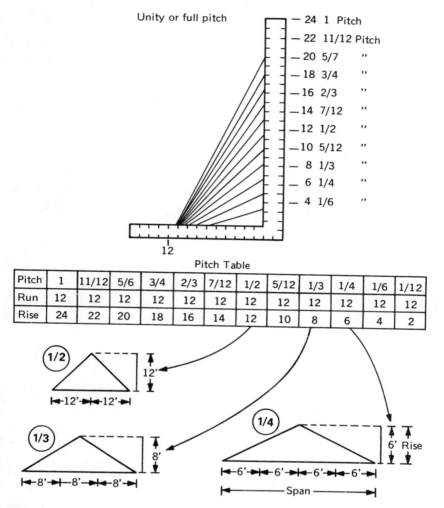

Figure 7-6   Pitch table, square and span.

Now look at the 1/3 pitch column in the pitch table. It shows a run of 12 and rise of 8. If the rise is 1/3 of the span and the rise is 8 ft, the span must be 24 ft. The pictorial beneath the pitch table shows this. The rise is indeed one-third the total span. Therefore, the mathematics is correct.

$$\frac{8\text{-ft rise}}{24\text{-ft. span}} = \frac{1}{3}\text{pitch}$$

The last picture illustrates the 1/4 pitch. Notice that the rise is indeed one-fourth the span.

$$\frac{\text{6-ft. rise}}{\text{24-ft. span}} = \frac{1}{4}\text{pitch}$$

Not all pitches need to be a fraction of a full pitch. Figure 7-7 illustrates that a pitch may indeed be a multiple of a unity or full pitch. The number 24 in the figure has been selected to simplify the understanding. A full pitch requires a span and rise equal to each other. A double pitch requires a rise twice the value of a span. If using Figure 7-7 as an example, a rise of 48 ft and span of 24 ft result in a double pitch.

$$\frac{\text{48-ft. rise}}{\text{24-ft. span}} = 2\text{ pitch}$$

Figure 7-7 Steeple pitches.

Problem: Suppose the building span was 10 ft and a double pitch was to be designed and built. What would its rise be in feet?

Transposing the formula: $\dfrac{\text{rise}}{\text{span}} = \text{pitch}$

To find rise: pitch × span = rise

Thus   2 × 10 ft = 20 ft

The total rise of the roof would be 20 ft. Pictorially the layout would look as in Figure 7-8. For each foot of run the rise is 4 ft. Since the run is one-half the span (5 ft) the total rise is 20 ft, as the math showed.

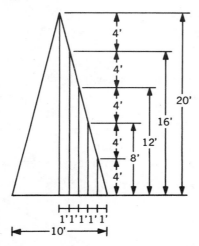

Figure 7-8 Quadruple pitch layout showing run and rise.

Refer once again to Figure 7-7 and notice that a pitch may also be triple or quadruple. These very steep pitches are frequently used on steeples of churches, and on the lower section of the gambrel or mansard roof design. The steep incline on contemporary structures such as townhouses is now referred to as a wall whenever its angle approaches 90 degrees from the horizontal.

Figure 7-9 Pitch notation on elevation drawings.

The current use of pitch in blueprints and elevation drawings is shown in Figure 7-9. It is a triangle located just above one of the sloping roof lines. The one shown indicates that the rise is 4 in. for each 12 in. of run, and this is commonly stated as "The roof has a pitch of 4 in 12-in. If there is more than one pitch, as in the gambrel roof design for example, two triangles would be included in the elevation drawings.

TERMS USED WITH THE COMMON RAFTER AND THEIR LOCATIONS

There are several terms used in describing parts or special cuts on a common rafter. Each one contributes to understanding some part of the rafter and is part of the terminology of carpentry. They are all shown in Figure 7-10 and defined below.

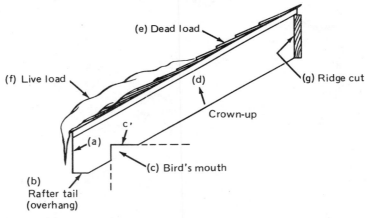

Figure 7-10 Common rafter terms and their locations.

1. The *overhang* or *rafter tail* is the section of rafter that extends out beyond the exterior wall line. This part of the rafter is used as part of the cornice of the building. The vertical cut (a), and many times the horizontal cut (b), are made on the end of the rafter tail. The rafter tail may project any distance past the wall; however, the most common overhangs are 6, 12, and 24 in.
2. The *birdsmouth* (c) is a right angle notch that is cut in the underside of a rafter. It is designed to allow the rafter to

seat properly on the wall. The horizontal cut (c'), also named the heel, seat, or bottom cut must be accurately made because the rafter and roof loads bear on the wall through this contact surface.

3. The *crown* (d) in a rafter is the natural bend through its length when sighted along the narrow or 2-in. edge. The use of the crown is always an important part in rafter layout and roof construction. All crowns must be set toward the sky or "crown up" so that the roof dead and live loads are borne better.

4. A *dead load* (e) is not exclusively an element of a rafter. It is the accumulated weight of the rafter plus the roof materials resting on it. These materials are the roof sheathing, tar paper and shingles.

5. The *live load* (f) is the weight placed upon a roof by the workmen, snow, ice, or the like resting on the roof for short periods.

6. The *ridge cut* (g) is a vertical cut at the top of the rafter. This cut must be made so that the entire surface attaches against the ridge board.

FORCE OF RAFTER AND LOADS PUSHING AT THE SUPPORTING WALL

The roof, especially the gable roof which is made from common rafters, develops considerable force unless proper steps are taken to reduce it. The weight of the rafters and other roofing materials try to settle straight down; but because they are A-shaped and fastened at the plates of the wall the force is directed to the wall. Figure 7-11 illustrates the idea just expressed as well as the final outcome if allowed to happen.

There are three methods that are used to overcome this condition. A *collar beam* or *collar tie* as shown in Figure 7-12a is nailed horizontal to the rafters and about 24 in. down from the ridge. This piece (1 × 4 or 1 × 6) is nailed to every other set of common rafters or a maximum of 48 in. on center. The result is that the ridge is held firmly in place and the loads on the roof are partially absorbed by the collar beam. The remainder of the loads are concentrated on the wall plate.

Figure 7-11   Results of forces when roof is improperly braced.

Figure 7-12   Bracing techniques—collarbeam, spanning brace, and truss bracing.

The second method used to overcome the remaining force is the use of ceiling ties between outer walls. L-shaped assemblies (Figure 7-12b) made from 2 X 4s and 2 X 6s are nailed together and to opposite plates before the rafters are installed.

The third method is the use of a truss brace shown in Figure 7-12c. It is installed from bearing wall midway through the building span at a 45-degree angle to rafters. Most of the time collar beams are sufficient, especially if ceiling joists are well nailed to the plates and each other. However, if especially large rafters are used and heavy deposits of snow are the rule rather than the exception, a combination of the three methods is recommended.

PROPER MEASUREMENT POINTS FOR DETERMINING RAFTER LENGTH

Thus far in this chapter the discussions and descriptions have centered on the roof assembly using the common rafter. This section begins the study on the rafter design. When rafters were beams hewn from logs, the measure of the rafter length was taken through the center of the width. It was understood that this measuring technique provided a more reliable length since edge widths were not always parallel and consistent. Today, machined beams and rafters are uniformly sized. This means that the crown edge of the rafter is the line to use in determining rafter length. The use of the top edge in no way alters the basic principle of determining rafter length.

To illustrate the proper points of measurement, and at the same time verify the comment made in the preceding paragraph, carefully study Figure 7-13. One of the first requirements in determining the rafter length is to determine the *run*. This is equal to one half of the *span* as shown in Figure 7-13. *A* in Figure 7-13 is a dashed line beginning at the outer edge of the wall and extending the full run. The other dashed line represents the length of the rafter without considering any rafter tail. The rise is measured as 6 ft. Since the run is 12 ft and rise is 6 ft, these measurements do correspond to the pitch of *6 in 12*.

Now note carefully that if a perpendicular line is drawn from the outer wall line up through the area of the rafter width to its top edge, a second set of lines (solid) can be drawn. These lines parallel the dashed lines. The result is that the rise is not changed; it still is 6 ft.

Figure 7-13   Laying out the rafter from its proper points of measure.

Two things are revealed: (1) the rafter length can be measured along the rafter's top edge; and (2) the rafter length is the longest line of the *right triangle* and is called the *hypotenuse* in mathematics.

The importance of knowing that the rafter length is equal to the hypotenuse of a right triangle is brought out dramatically during the estimating phase of bidding on the construction job. By sitting at the desk the carpenter-builder is able to calculate the rafter length to determine what size lumber they will be cut from. How does he do this? The formula that must be used was developed as the *Pythagorean theorem* centuries ago. It looks like this:

$$ac = \sqrt{ab^2 + bc^2}$$

Translated to rafter terms it looks like this:

$$\text{Rafter length}[1] = \sqrt{\text{run}^2 + \text{rise}^2}$$

[1] Does not include overhang or rafter tail or reduction for one-half the ridge thickness adjustments.

To calculate the rafter length from values given in Figure 7-13, the problem looks like this:

$$\text{Rafter length} = \sqrt{\text{run}^2 + \text{rise}^2}$$
$$= \sqrt{12^2 + 6^2}$$
$$= \sqrt{144 + 36}$$
$$= \sqrt{180}$$
$$= 13.4 \text{ ft. or } 13 \text{ ft 3 in.}$$

Now here is where a calculator comes in handy.

1. Square the length of the run and store it in memory or write it down.
2. Square the rise after first calculating it.[2]
3. Add the squared value of the run to the squared value of the rise and use the square root button ($\sqrt{}$) to find the rafter length.
4. Convert the decimal portion of the result to inches.

The procedure used on most calculators goes like this:

run = 12 ft

$\boxed{1}\,\boxed{2}$ $\boxed{F}$ $\boxed{X^2}$ STO or write on paper

rise = 6 ft

$\boxed{6}\,\boxed{F}\,\boxed{X^2}\,\boxed{=}\,\boxed{+}\,\boxed{RCL}$ (or load 144 manually) $\boxed{=}$

$\boxed{F}\,\boxed{\sqrt{}}$ and you have the answer.

Some calculators that have function buttons for X^2 and do not need the step F.

With the rafter length known, any overhang can be added to obtain its full length. This length of the rafter may be retained for use on the job where it can be measured with a tape. However, the carpenter's square is frequently used.

A well-designed carpenter's square provides tables on one of its surfaces for the hypotenuse values per foot of run for many different pitches. Before examining the variety of them, an

[2] Total rise = rise per foot run × no. of full and partial foot run; e.g., 5 in. × 10 ft. 6 in. = 5 in. × 10½ ft. = 52½ in. rise.

understanding of their origin and accuracy is important to know. To help illustrate this subject, Figure 7-14 is provided.

Figure 7-14 Framing square numbers correspond to the length of the rafter for each foot of run.

For a rafter run of 10 ft there is a 5 in. rise per foot for a total rise of 4 ft 2 in. (50 in.). Mathematically, the hypotenuse of each 12 in. is 13 in.

$$\text{rafter length per foot of run} = \sqrt{\text{run}^2 + \text{rise}^2}$$
$$= \sqrt{12^2 + 5^2}$$
$$= \sqrt{144 + 25}$$
$$= \sqrt{169}$$
$$= 13.00$$

Notice the partial table found at the top of Figure 7-14. Tables of this kind are found on many different types of framing squares. Under the 5-in. mark is the figure 13.00. This is the value that was found by calculations. If a 6 in 12 pitch were used, the value 13.42 under the 6 is the rafter length for each 12 in. of run of a common rafter.

To assist carpenters and builders with the calculation of rafter lengths, especially during estimating and even on the job, the table and accompanying drawings in Figure 7-15 provide lengths per foot run of the most common pitches.

Table for Length of common rafters per foot run

Rise	3"	4"	5"	6"	7"	8"
L/ft run (in)	12.37	12.65	13.00	13.42	13.89	14.42
Rise	9"	10"	12"	15"	18"	24"
L/ft run (in)	15.00	15.62	16.97	19.21	21.63	26.83

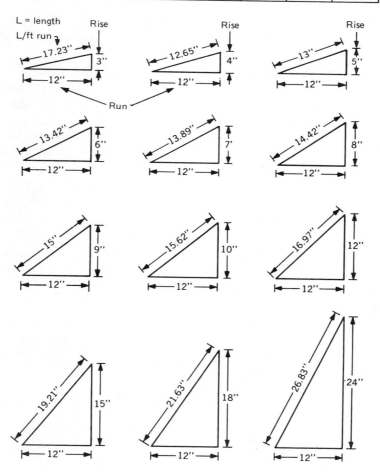

Figure 7-15 Quick reference table and illustrations for obtaining the rafter length per foot run for all commonly used rises.

COMMON RAFTER LAYOUT, PHYSICAL POSITIONING

Review Figure 7-16 until the following data have been found:

1. The building span line
2. The rafter run line
3. The rafter rise line
4. The rafter slope line
5. The 2 x 4 block that represents the wall's plate.

Objective: Simulate a rafter's position by using the wall (or driveway) as a reference area and formally lay out and fit a rafter to the layout.

A plate-reference chalkline needs to be snapped from 12 to 18 in. above the sill of the wall. This allows sufficient height for the projection of the portion of the rafter that will overhang the wall line and become the rafter tail.

Calculate the total rise of the rafter by using the formula: rise X rafter run = total rise in inches. The line for the rise must be perpendicular to the baseline so a plumb bob can be used effectively. Measure the rafter run along the baseline. Suspend the plumb bob above the mark on the baseline and mark the upper end of the plumb bob's line. Snap a line connecting both points. If the total rise is greater than the height of the wall, temporarily nail a 1 X 6 or 2 X 6 approximately centered where the vertical line is to be snapped. Snap the line on the board. Measure *up* from the baseline and mark the total rise. Then snap the slope (hypotenuse) lines between the total rise mark and the outer corners of the baseline. Next, nail a scrap piece of 2 X 4 X 12 in. alongside the corner stud with the block's *top* flush with the base chalkline. Drive a 16d common nail partially into the 1 X 6 or 2 X 6 (if used) at the intersect point of the vertical span (center) line and the slope line.

Lay a rafter on the nail and 2 X 4 block and, after ensuring that the ridge end is fully beyond the center line, tack the rafter in place with an 8d common nail.

Laying Out a Ridge Cut

Lay out the ridge cut, using the level shown in Figure 7-16b. Next, subtract one-half the thickness of the ridge board by measuring in a distance from the vertical line just drawn. Draw another vertical line even with the mark just made and label it "cutting line." While the rafter is in position, the bird's mouth and

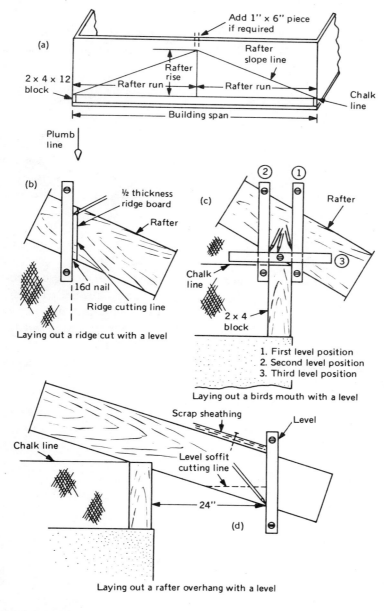

Figure 7-16 Common rafter layout—physical position.

overhang should be laid out, after which all cuts to the rafter can be made.

Marking a Bird's Mouth

Use a 24-in. level and a pencil to make a bird's mouth layout as shown in Figure 7-16c. This technique results in a seat equal to the 2 × 4 stud width.

Laying Out an Overhang or Rafter Tail

Only one example is provided of the laying out of an overhang. You must use the principle to fit the specific requirements. Figure 7-16d shows how the principle is employed. Assume that a 24-in. overhang is being laid out. Measure on a horizontal plane (from the 2 × 4 block) 24 in. and make a mark. Use the level in plumb fashion to make a vertical line.

Next, review the plans and determine the width of the fascia board so that the amount of the lower edge of the rafter end that needs to be trimmed is defined. Also note how much is needed and used to retain the soffit if a box cornice is installed. The plan should show this detail, or it may be calculated as follows: Subtract the lip and the soffit's thickness from the total width of the fascia board. Record the remainder.

Lay a piece of roof sheathing material (a scrap piece) on the rafter's top edge, and with a ruler measure down from the plywood's top edge a distance equal to that recorded. Mark along the vertical line previously drawn. Draw a level line from this mark *back* toward the bird's mouth. The rafter-end layout is complete.

Checking the Cuts and Verifying the Fit

Cut out the rafter and another one. Try them for fit on the opposite sides with a scrap piece of stock that represents the ridge inserted. If they fit (which they should), use the rafter as a pattern to lay out and cut all rafters required. **Caution** Be sure to keep the crown *up* on all rafters.

Common Rafter Layout, Stepping-Off Method

Figure 7-17 Common rafter layout—stepping-off method.

The stepping-off method is a step and repeat technique that is frequently used to step-off the rafter length in successive units. Briefly, the process involves a move or step of the framing square for each full foot of rafter run and a partial step for a fraction of a foot of run. The connecting points of framing square to rafter's top edge are the pitch numbers. In Figure 7-17 points A and B are the pitch numbers that are found on the square and used along the rafter's top edge. At point A the 4 from "4 in 12" is used; at point B the 12 is used. Since the run is 6 ft 4 in., six full steps are used. While the square is positioned for the sixth step, a mark should be made on the rafter where 16 in. falls on the square (see section c). The seventh step, a partial step, is made using 4 in 12 but the square is aligned over the 16-in. point marked during the sixth step. The ridge and tail cuts are perpendicular and the bird's mouth is made as Figure 7-17b shows. The seat cut should always be equal to the 2 × 4 plate width or wider.

The stepping-off method, although used extensively, is not accurate. If each step of the square is just a pencil's thickness off, the rafter's length could be one-sixteenth of an inch longer or shorter per step. An experienced carpenter is very careful to avoid this situation by using one edge of the pencil mark rather than the entire mark. By using the edge of the pencil mark greater accuracy is achieved throughout the stepping-off sequence.

Observation

Much time and effort can be saved in estimating rafter needs and laying out rafters if a calculator is used to do the math necessary to determine its length. The formula is very easy to use once it is understood. The carpenter must learn how to change a decimal into a fraction of an inch, but a short table can be made for quick reference. When the construction industry switches to metric measurements calculations will become very easy because no conversion to fractions is needed.

The different cuts needed to be made on a rafter can be done easily by first laying them out with a framing square. These include the ridge cut, bird's mouth cut, and rafter tail cut. Before making the ridge cut, an allowance for one-half the ridge board thickness must be subtracted from rafter length; then the cut can be made. The rafter tail cut can be laid out and cut initially, that is, when the rafter is on the bench, or it may be done after all rafters are installed. Many experienced carpenters believe that cutting the rafter tails after the rafters are in place makes for a more perfect building line. There is considerable truth in this idea, since frequently there may be variations in top plate alignment.

SELECTING THE PROPER SIZE RAFTER

The same engineering data that went into selecting the joists (Chapter 5) is used to select rafters. Tables in Appendix C summarize industry standards for rafters and their loads. Modulus of elasticity and fiber in bending factors are again primary for quality control of the lumber grade and species necessary for adequate support. Both live and dead predicted loads must be considered along with the slope of the roof. The usual calculated live loads result from snow deposits and personnel working on the roof. Obviously snow is more serious. Values of 20 psf, 30 psf, and 40 psf are consistently used as live-load values. Dead loads usually are established as 10 psf to 20 psf. The slope of the roof (pitch) is also considered as a factor. Some tables in the appendix are compiled for pitches of less than 3 in 12 while others are for pitches of high slope, or greater than 3 in 12.

Recall the three ratings for stiffness are $S/300$, $S/240$, $S/180$. It was stated in Chapter 5 that an $S/180$ rating would be acceptable for rafters under some conditions. $S/360$ and $S/240$ are both acceptable since the load stresses placed on floor joists are usually considerably more than on the roof. A study of the tables in Appendix C shows that certain members (block type) are limited to $S/180$ and others are limited to their grading.

The next two problems aid in understanding the importance of selecting the proper timbers for rafters.

Problem 1: Select a rafter to carry a 40 psf load whose pitch is 6 in 12 and will be used on a building with a 36-ft span. Spacing will be 24 in. o.c. and rafters must have a deflection limitation of $S/240$.

   Solution:

1. Rafter run = 18 ft or half the building span.
2. At a rise of 6 in 12 the rafter length (not including overhang) is 20 ft.
3. Only 2x 12 grade Dense Sel Str can be used. See Table 9, Appendix C.
4. Dense Sel Str has E of 1.9 million psi. F_b of 2100 to 2400 psi.

5. Species meeting both grade and load standards are Douglas fir and larch. See Appendix A.

Problem 2: Select a rafter by species and grade that will meet the following requirements. The building will have a span of 26 ft and an 8 in 12 rise. Predicted dead load will be 15 psf. Allowable unit stress in bending (F_b) will not exceed 1200 psi under normal conditions nor 1500 psi with a seven-day loading. Rafter spacing equals 24 in. o.c.

Solutions:

1. Rafter run = 13 ft or half the building span.
2. At a rise of 8 in 12 the rafter length (not including overhang) is 16.5 ft.
3. Several possible grade choices (see Table 7, Appendix C):
   a. 2 x 8 grade (Sel Str)
   b. 2 x 10 grade (No. 2)
   c. 2 x 12 grade (No. 3)
4. Species data for the three grades (see Appendix A):
   a. For 2x 8 rafters grade Sel Str, the acceptable species are:
      1. **Douglas fir, larch** (F_b normal 2050 psi; seven-day load 2560 psi)
      2. Douglas fir south (F_b normal 1950 psi; seven-day load 2440 psi)
      3. Southern pine (F_b normal 2050 psi, seven-day load 2560 psi)
      4. Eastern hemlock (F_b normal 1750 psi, seven-day load 2190 psi)
      5. Eastern spruce (F_b normal 1500 psi; seven-day load 1880 psi)
      6. Engelmann spruce (F_b normal 1350 psi; seven-day load 1690 psi)
      7. Hemlock fir (F_b normal 1650 psi; seven-day load 2060 psi) and several others.
   b. For 2 x 10 rafters No. 2, all the previous species are acceptable except:
      1. N/A Eastern spruce (F_b normal 1000 psi; seven-day load 1250 psi)

2. N/A Engelmann spruce (F_b normal 950 psi; seven
-day load 1190 psi)

c. For 2 x 12 rafters Grade No. 3, none of the previous
species are acceptable. Typical reason: N/A Douglas fir,
larch No. 3 has F_b normal 850 psi and seven-day loading 1060 psi.

5. Stock selection: Acceptable grade and species include: the
2 x 8, Grade *select structural* hemlock, fir or better; or
2 x 10 Grade No. 2 hemlock, fir, Douglas fir, or southern
pine.

The methods used to solve these two problems are based on
sound engineering data. All judgments of this kind should be
founded with facts of the types listed in the appendixes and taken
from the blueprints or specifications. Where a range of species and
sizes of materials is adequate to perform the functions of rafter,
economics becomes an important factor. Holding costs low is good
business sense as long as the quality of the materials for the job are
maintained properly.

PROPER LOCATION OF RAFTER ON OUTER WALL

Each rafter in a roof assembly needs to be spaced 16-in., 24-in.,
32-in., or 48-in. on-center depending upon the design of the roof.
The spacing does not, however, dictate each rafter's position on
the wall with respect to the stud position in the wall, although it
could. The roof load bears its weight onto the wall through the
rafters and it is important that the plates bear as little of this
weight as possible. Therefore the force of the roof's weight should
pass through the plates into the studs and to the foundation.

To provide the best opportunity for the transfer of force to
be passed to the studs on posts and not be borne by the plates, it
is essential that rafters be aligned either directly over the stud or
post or aligned adjacent to the stud. If rafters are installed midway
between the studs, the on-center spacing will cause the plates to
bear the load and the possibility of shearing the plates exists.

Several opportunities exist for the carpenter in the layout of
the rafter positions on a wall or each rafter aligned directly over
the stud or post in the wall as Figure 7-18a shows. This assures

that the roof load is borne directly by the studs or posts and transferred through the plates with no shear pressure. The ceiling joists are aligned next to the rafters and both are nailed to each other and fastened to the plates. Figure 7-18a is a front view of the isometric drawing of the wall and roof assemblies.

Figure 7-18   Rafter positioning on the outer wall—alignment with studs.

By adopting this plan for rafter positioning, the layout on wall plate and ridge (if used) begins at the same corner as the layout for wall studs and floor joists. This technique places all structural members in alignment.

If the alternative plan is used as Figure 7-18b shows, the ceiling joists are aligned over wall studs. The rafters are then located either side of the ceiling joists. This technique locates all rafters within 2 in. of the wall stud and is acceptable in building codes. There is, of course, some shear force placed upon the plates by the roof load. However, the plate stock does have sufficient strength to accept this load because of the close proximity of the stud alignment below it.

If beam-and-post construction is used, the rafter beam is aligned and fastened directly to the supporting post below it. This type of construction is used where exposed beam or cathedral ceilings are used.

PROBLEMS AND QUESTIONS

1. In which of the following roof designs may a common rafter be used?
   a. gable   b. hip   c. shed, gambrel   d. dormer
   e. all of the above
2. Which fractional value indicates the steepest pitched roof?
   a. 1/4   b. 1/3   c. 1/2
3. When defining a pitch is the *run* or *span* of a building used along with the rise?
4. What is the pitch of a roof of a building that has a span of 12 ft and rafter rise of 6 ft?
   a. 1/4   b. 1/2   c. 3/4
5. Transpose the formula *Rise/span = pitch* to find the rise when span and pitch are shown.
6. What does the symbol below mean on an elevation blueprint?

7. How is the notation in the symbol above read?
   a.   4 on 12   b.   4 in 12
8. What portion of a common rafter would have a *rafter tail*?
   a.   The lower portion
   b.   The portion near the ridge cut
   c.   Rafters never have tails

9. What is the purpose of a *bird's mouth*?
10. Define the term *crown* as it applies to a rafter.
11. What parts of a roof account for the dead load on a rafter?
12. Why is the ridge cut always a plumb cut?
13. What member added to a gable roof reduces the forces that are on the plates?
    a. rafter   b. collar beam
14. List two methods that prevent the roof from causing the walls to bow outward because of forces.
15. Why can a uniformly sized rafter be measured for length along its crown edge?
16. What is the length of a common rafter if its run is 13 ft 6 in. and the rise is 6 in 12?
    a. 13 ft 6 in.   b. 15 ft. 1/8 in.   c. 17 ft 2-1/2 in.
17. What is the lowest grade and smallest size of lumber appropriate for a rafter that is to run 12 ft, have a rise of 7 in 12, and have a deflection not to exceed $S/240$.
18. Why is it important to align rafters over wall studs?

8

The Roof, Part II:
Equal-Pitch Hip and Valley Rafters

Hip and valley rafters are more difficult to construct than the common rafter because more variables are involved in their calculations and fitting. Defining their location is relatively easy; however, understanding how to obtain their length requires fundamentals of rise, run, the pythagorean theorem (right triangles) and adjustments to the top edge of the rafter and bird's mouth.

The method of determining a rafter's length in this chapter uses the approach that the hip or valley is framed in a balanced or equal pitch design. This means that the left and right portions of the hip or valley are of equal length.

Selecting the proper rafter size is important for several reasons. They are usually made from stock at least 2 in. wider than common rafters so that jack rafter bias cuts can have continuous surface contact. The load-bearing factors also play an important part in the rafter selection process.

DEFINING THE LOCATION OF HIP AND VALLEY RAFTERS

Figure 8-1 illustrates a house that has a roof framed with four hip rafters. Three of the four hips are shown in this figure. Notice that the lower edge of the hip centers over the corner of the building.

In Figure 8-2 a house with a roof framed with both hip and valley rafters is illustrated. As in Figure 8-1, the hip rafters of the house in Figure 8-2 have their lower edges in line with the corners. The hip rafters in this figure are designed to be of equal pitch. There is one valley rafter in this roof, forming an interior corner. Its lower edge aligns with the interior wall corner.

Figure 8-1   Locating the position of hip rafters in a hip roof.

Figure 8-2   Locating hip and valley rafter in a hip and valley roof.

Each dormer on the front of the house shown in Figure 8-3 is framed with two valley rafters. These are located at the intersect point of dormer roof to the main roof. These valley rafters also employ the regular valley design, but may sometimes use the *unequal pitch* design principle.

Figure 8-3   Locating valley rafters in a roof and dormer.

TERMS USED WITH HIP AND VALLEY RAFTERS

There are several terms used with hip and valley rafters that are the same as those used with common rafters, and there are several special terms and adaptations of common rafter terms. The terms each have in common are the *rafter rise* (in inches) and the *rafter run* (although the run per foot will not be 12 in.). Calculation of rafter length can be made using the pythagorean theorem. The hip or valley rafter requires a *bird's mouth* cut, *ridge* cut, and may need a *tail* or *overhang* with tail cut.

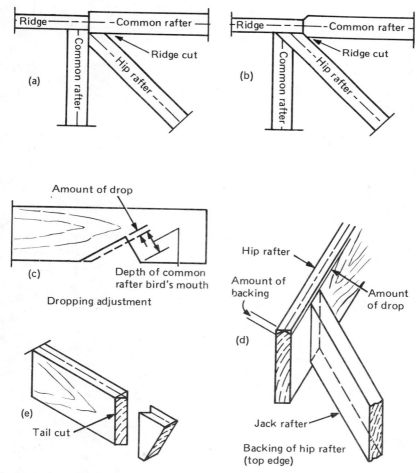

Figure 8-4 Defining terms and their location on a hip rafter (they also apply to valley rafters).

The new and adapted terms are shown in Figure 8-4. The plumb cut of the hip and valley that intersects the ridge and common rafters must be altered with bevel or side cuts as Figures 8-4a and b show. At *a* its length is shortened and two angle cuts are made along the plumb cuts. In *b* an adjustment for length is made and single angle cut is made so that the hip or valley rafter butts against the ridge. In Figure 8-4c a new term *drop* or *dropping adjustment* is shown. This is an adjustment cut made in the bird's mouth to lower the rafter to the level of common and jack rafters. If dropping adjustments are not used, then *rafter backing* must be used as Figure 8-4d shows. (Valley backing is shown later.) The top or crown side of the entire length of the hip must be backed with a bevel cut so that jack rafters intersect properly and roof line at the hip is proper. Finally, the *tail cut* Figure 8-4e shows is plumb cut with two bevel cuts; each cut surface must form an angle that aligns with the tail cut of jack and common rafters.

DETERMINATION OF HIP OR VALLEY RAFTER PITCH, RISE, RUN, AND LENGTH

Pitch

A review of the elevation plans shows the roof line will quickly identify the pitch of a roof. In the case of a hip or hip and valley roof or just valleys in a valley and gable roof more than one pitch may be indicated. If only one is shown then all roofs are sloped at the same angle; that is, they all rise to the same height and their runs are equal.

Constructing a roof with a single pitch is easier to understand and build than one having several pitches. An examination of a *regular* or *equal pitch* roof is used that has a pitch of 6 in 12.

Defining the Position of a Hip Rafter in an Equal-Pitch Roof

The first requirement is to obtain the outer dimensions of the building, which include its span and length. The span should be a measurement that is shorter than the length. Assume that a building with lines of span and length as shown in Figure 8-5 is to have a hip roof installed. With a building span of 28 ft 0 in. the run is 14 ft 0 in. Since the roof pitches of the span and length are equal (6 in 12), a like distance of 14 ft 0 in. is laid off between points B and D and E and F. A line is drawn connecting points D and F. Midway, or 14 ft 0 in. from either D or F point A is defined. This establishes the hip rafter positions as AB and AE. Since squares were established, the hip rafter is at 45-degree angle to the span and length of the building line. The same method can be used to define the position of a valley rafter.

Figure 8-5 Defining the hip rafter run (also applies to valley rafter run).

Defining the Run of a Hip or Valley Rafter

Using Figure 8-5 again as an example, the hip rafter run is defined as a distance measured from points A to B on a level line (not up the slope of the rafter). Two methods are used for finding the length of run.

Solution 1: Using the pythagorean theorem..

Hip rafter run $= \sqrt{BC^2 + AD^2}$

$\qquad = \sqrt{14^2 + 14^2}$

$\qquad = \sqrt{196 + 196}$

$\qquad = \sqrt{392}$

$\qquad = 19.7989$

Hip rafter run $= 19.7989$ ft $= 19$ ft 9 5/8 in.

Solution 2: Using tabulated data from Figure 7-15, a rise of 12 in. and a run of 12 in. result in a hypotenuse length of 16.97 in., which is the length of hip run per 12 in. of common rafter run. Therefore,

Hip rafter run = length per 12 in. times the number of ft of common rafter run

= 16.97 in. × 14

= 237.58 in. ÷ 12   (divide by 12 to obtain answer in ft)

= 19.7989 ft = 19 ft 9 5/8 in.

This means that for each 12 in. of common rafter run the hip rafter (and valley rafter) has a run of 16.97 in. or roughly *17* in.

Defining the Hip or Valley Rafter Projected Length

The framing square has a table imprinted on it that provides the length per foot run of equal-pitch hip or valley rafters for many different rises. It uses a basis of 16.97 in. as a base leg or *run*. Figure 8-6, which includes Table 8-1, provides the length per foot of run for the usual rises of roofs. In the example problem above, the pitch was established as 6 in 12. From Figure 8-6, a rise of 6 in 12 for a hip or valley rafter has a *length/ft* run of 18 in. Therefore, the rafter length may be calculated:

Hip rafter length = length per foot run × no. of ft of common rafter run

= 18 in × 14

= 252 in ÷ 12   (to obtain the number of feet)

Hip rafter length = 21 ft

With the rafter length calculated it becomes important that its measurement be properly established on the rafter itself. Recall that dimensions for rafters may be made through the center of the rafter or in the case of sized members along their top edge. When using the top edge a perpendicular line (plumb line) must be used to ensure no change to rafter length is made (see Figure 7-13).

Table 8-1 Length of hip or valley rafter per foot run

Rise	3"	4"	5"	6"	7"	8"
L/ft run (in)	17.23	17.44	17.69	18.00	18.36	18.76
Rise	9"	10"	12"	15"	18"	24"
L/ft run (in)	19.21	19.70	20.78	22.65	24.74	29.39

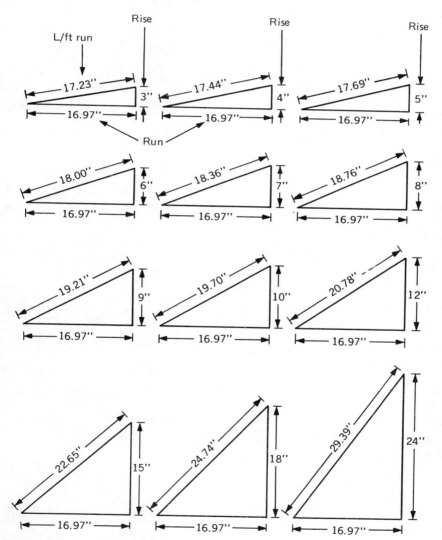

Figure 8-6   Length of hip or valley rafter per foot run.

Figure 8-7 Laying out the hip rafter from the rafter centerline.

180

The rafter is laid out as shown in Figure 8-7. Notice that the ridge (plumb) cut is made using 17 in. and 6 in. on the square as is the bird's mouth.

Up to this point the only difference between layout on a common rafter and hip or valley rafter is using 17 in. instead of 12 in. on the square to obtain the plumb and horizontal cuts. However, the remaining cuts need to be made from a new reference point—the center of the top edge of the rafter. Notice in the top drawing in Figure 8-7 that hip rafter's length is from the outer corner of the wall to the midpoint (intersect) of the junction of the common rafters and ridge. This means that some special cuts must be made to each end of the rafter as well as an allowance for length near the ridge.

The lower section of Figure 8-7 illustrates the side view of the hip rafter with its primary cuts for ridge, bird's mouth, and lower end. These may all be laid out with the framing square or with the use of a miter or bevel square (this method will be shown later). Also notice the top view of the rafter with the broken line bisecting the rafter thickness. This is the center line that must be used as a reference for the remaining cuts.

Rafter Cuts and Adjustments

Lower-edge bevels. The lower-edge bevels are made at the angle back from the center line and are cut accordingly. These surfaces then align with the wall plates or sheathing. In the lower end of the valley rafter, the bevels are made opposite those of the hip as Figure 8-7c shows. The layout point is still the centerline of the rafter. These bevel cuts also align with the wall plates. It would appear that simple 45-degree angles should be cut beginning at the centerline. If the rafter were laid on the horizontal plane this idea would be accurate; especially so since the layout of the 45-degree lines are made on the top surface of the rafter. However, when the rafter is tilted up as each rise predicts it will be, these angles must be cut differently so that their cut surfaces still remain in alignment with the wall plates.

The concept just stated is probably one of the most difficult to understand. Figure 8-8 may provide the visual reinforcement needed to understand why the layout and side cuts of the hip and valley are not 45-degree angles. To begin, the combination square

An angle less than 45°

Miter or bevel square

Hip rafter

90 degrees

Horizontal plane

Combination square

Wall top plates

Figure 8-8 Bevel miter cuts align with wall plates, miter square angle is less than 45 degrees to achieve the miter needed.

182

(a)

Hip or velley rafter; marking the side cut

Rafter center line

Bevel square

(b)

Hip rafter; marking the side and tail cuts

(c)

Valley rafter; marking the tail cut

Figure 8-9 Marking hip and valley side cuts and tail cuts with properly set miter square.

is held at a horizontal plane with the plates (see X). In this position the beveled edges of the hip rafter form a 90-degree corner, 45 degrees on each half. However, the miter or bevel square shown resting on the top surface of the rafter is not on a horizontal plane with the plates since the hip rafter is shown at some rise. Because of the position of the bevel square and hip (or valley) rafter the angle of the bevel square is *always* less than 45 degrees when positioned as shown.

Most of the better quality framing squares have a table on the blade that gives the setting for hip and valley side cuts for many different rises. Once understood, they may be applied as needed. However, there are several other methods that may be used. One is to use the bevel or miter square. Setting it for the proper angle is easy. Figure 8-9a shows how it is used to make a single side cut for hip or valley where the rafter ties into the ridge or other valley. In Figure 8-9b side cuts for ridge and lower-end tail cuts are shown easily laid out with miter square. Finally, in Figure 8-9c inside bevels are laid out with miter square for the lower end of the valley rafter. These illustrations lead to the obvious question, "How does one obtain the angle needed for a particular rise?"

Table 8-2 and Figure 8-10 combine to provide visual understanding as well as a standard of values to lay out side cuts for almost all different rises. Table 8-2 provides the following data:

1. Rises begin at 3 in 17 up to 18 in 17.
2. Degrees of side cut for each different rise.
3. A conversion from degrees to a numerical value stated in inches. The 2 stands for the framing square's 2-in. blade. The 2-1/16L across from the 3-in. rise stands for a base leg of just under 2-1/16 in. (The L stands for light or shy measurement).
4. The last column provides bevel square settings using the framing square's blade as a basis.

Now we will apply the tabulated data to a roof with a 6-in. rise. As Figure 8-10 shows, the bevel square is set so that its wooden surface is against the outer surface of the framing square.

Its metal blade is set to the values listed in column 4 of Table 8-2, "1 in. out, 1-5/8 in. inside." When the two values are in alignment with the bevel square's blade, lock the blade in position.

The lower portion of Figure 8-10 shows an expanded inch scale with each of the rises from Table 8-2 located for ease of reading. Any serious carpenter who uses this method of laying out side cuts on hips and valleys could easily duplicate this expanded segment on a small card and carry it in his wallet for quick, reliable reference.

Ridge adjustment and bevels. Depending on the technique of construction used, the ridge cut on a hip (or valley) rafter may need to be adjusted for length. Figures 8-11a and a' show two techniques for making the ridge cut on hip rafters. In a the rafter length must be shortened equal to the distance between points *A* and *B*. This is necessary for several reasons. First, the rafter length was originally calculated from point *A* on the ridge to the outer corner of the wall. (Recall this was 21 ft 0 in. in the example). Since the common rafters both center on point *A* and are installed as shown, a portion of their surface accounts for some reduction of hip rafter length. Finally, a portion of the ridge board's thickness must be subtracted from the rafter's length. (pp. 188-89)

TABLE 8-2   Hip and Valley Side Cuts with Bevel Square Settings

Rise	Angle (Degrees)	X (in.) = 2/Tan	Bevel square setting on framing square (see Fig. 8-10)	
3	44.5	2-1/16L	1-in. Out	1-9/16L
4	44.2	2-1/16		1/9/16
5	43.8	2-1/16F		1-9/16F
6	43.3	2-1/8		1-5/8
7	42.7	2-3/16L		1-11/16L
8	42.2	2-3/16F		1-11/16F
9	41.5	2-1/4		1-3/4
10	40.8	2-5/16		1-13/16
11	40.2	2-3/8		1-7/8
12	39.5	2-7/16		1-15/16
15	37.0	2-5/8F		2-1/8F
18	34.5	2-15/16		2-7/16

Note: L behind number stands for *light* and F for *full*.

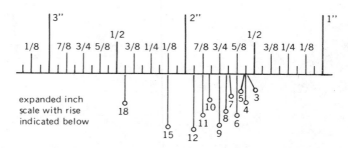

Figure 8-10   Miter square settings for various side and tail cuts for various pitches.

Two practical methods of defining the adjustment length are (1) measuring the distance from point *A* to point *B* where it intersects the corner of the common rafters at the ridge, and (2) lay out a full-scale model of the ridge and common rafters on a scrap board or cardboard and then measure the distance from points *A* to *B*.

After determining the adjustment length, mark the hip rafter accordingly along its top center line; then lay out the required bevels and cut along the markings.

An alternative technique of fitting the hip ridge cut is shown in Figure 8-11a'. In this technique the hip rafter is butted to the ridge rather than the common rafters. The adjustment for length is made on the rafter as shown; however, the bevel cuts result in one side cut across the full hip rafter with point B as the intersect point of the rafter's centerline and ridge board edge.

While studying this technique, notice that several adjustments must be made to the common rafters. Common rafter 1 must have both of its ridge cut sides beveled, and common rafter 2 must have its ridge cut made at an angle equal to that of a hip jack rafter (covered in Chapter 9).

The valley rafters shown in Figure 8-11b and b' also require adjustment of their initial lengths to account for ridge board thicknesses. In b, where both ridges of the valley are of equal height, the adjustment is made from point C to D. Then bevel cuts are made from the adjustment point D when it intersects the middle of the rafter thickness. Figure 8-11b' shows that one valley rafter (left) intersects with the main roof ridge, and the second valley rafter intersects with the lower ridge and first valley rafter. The adjustment for length for one valley rafter is the distance C to D. The adjustments and bevel cuts in Figure 8-11b' are taken between points C to D and E to F, respectively. All of these angles for side cuts can be laid out using the bevel square, and the plumb cuts should be made using the framing square and the rise and 17 in.

Backing and Dropping. With the rafter adjusted for length and beveled to fit either between common rafters or against the ridge, it would seem appropriate to install it. Not so, a *backing cut* or *drop cut* must be made first. Figure 8-12a illustrates backing on a hip rafter. Imagine that a sighting is made along the top corner of all the common rafters and jack rafters to the hip rafter. All surfaces are in line along the level line except the hip; it is higher. Notice also that accurately cut and installed jack rafters (the subject of Chapter 9) are short when placed against the unbacked hip rafter; that is, they appear to be short because they do not reach the lip of the hip. This area above the level line and from jack rafter to hip rafter lip is the portion of the rafter to be backed. (page 190.)

(a)

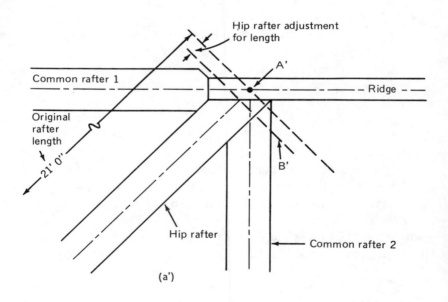

(a')

Figure 8-11   Ridge adjustment layout on hip and valley rafter and cornice.

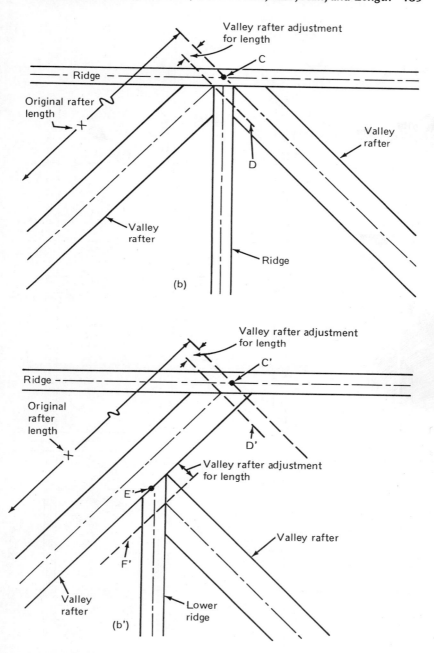

(b)

(b')

Figure 8-11 *(continued)*

(a)
Hip rafter
without backing cut

Backing
area

Top of jack
rafter and
hip contact
point

Jack
rafter

Level line

Bevel
cut on
hip

Wall plates

(b)
Misalignment of
sheathing due to
not backing
hip rafter

Level line

Wall plates

(c)
Hip rafter
with backing cut

Top of jack
rafter in
line with
backed hip

Level line

Figure 8-12   Backing the hip rafter to allow proper alignment of sheathing.

Before examining how to do the backing, consider what happens when sheathing is installed on a hip roof where the hip rafter is unbacked. As Figure 8-12b shows, the sheathing is lifted and only makes contact with the lip of the hip. The space under the sheathing on the hip is of little consequence since nails may easily be toenailed at a slight angle into the hip rafter. However, the roof line is not proper and this is the problem. Cornice construction as well as roof shingling are both affected. Backing eliminates the problem, as Figure 8-12c illustrates. Notice that one-half of the top surface of the hip rafter is now flush with the level line. Notice also that because of the backing cut the jack rafter's top edge intersects flush with the backed surface of the hip rafter. If sheathing is installed now it lies flat and has a flat contact surface on the hip. Cornice construction can now be made accurately and shingles lay properly on the roof.

Figure 8-12 only shows half of the hip rafter backed surface; what does a fully backed hip rafter look like and how is the backing cut laid out? Figure 8-13 is provided to aid in understanding. At the left of the figure is an unbacked hip rafter's lower end. Notice that the plumb cut is beveled at horizontal 45-degree angle back from the center point. To its right is the same hip rafter after backing has been made. Notice the correlation of the right side backed edge to the level line used in Figure 8-12.

Finally, look at the right of the figure and observe the framing square and a section of hip rafter. To lay out the backing place the square as shown with 17 (the run) and the rise (in this example 6 in.) along the lip of the rafter's top edge. If the hip rafter is of nominal 2-in. stock (actual thickness 1-1/2 in.) divide by 2 and use the results of division as shown to obtain the dimension of backing. One-half of 1-1/2 in. is 3/4 in., therefore the backing line is established at the 16-1/4-in. mark on the framing square. For a 2 x__hip rafter with a 6-in. rise the backing cutting line is 1/4 in. from the lip of the rafter.

This method, once learned, is easily repeated whenever the need arises, but a reference to the amount of backing needed for all rises reduces the work involved. Table 8-3 provides a quick reference for the amount of backing needed for the usual rises of roofs. Those given are for a nominal 2-in. thick hip rafter. Note, however, that if a nominal 1-in. thick hip rafter is used all values in the table should be reduced by one-half.

Laying out the backing
or drop (nominal 2-in.
thick rafter)

* See table 8-3 for all values of backing or drop

Figure 8-13   Backing layout with a framing square.

Setting a power saw to cut the backing after determining and laying it out is time consuming but correct. Frequently, where roofs are pitched up to 8 in 12 and sometimes even 10 in 12 a *drop* is made in lieu of the backing. It is quicker to make. Both Figure 8-13 and Table 8-3 indicate the amount of backing is equal to the amount of drop. Where then is the drop cut made?

A *drop cut* is made on the seat cut of the bird's mouth as in Figure 8-14. If, for example, a drop cut is used on a hip (or valley) rafter, with a pitch of 4 in 12 the amount of drop is 3/16 in., as Table 8-3 indicates. A line is drawn parallel 3/16 in. from the bird's mouth common rafter cut, and then this stock (blackened area) is cut away. When the rafter is installed it is lowered 3/16 in., thereby bringing the top lip level with the jack and common rafters. A small gap will exist between sheathing and rafter top surface, but this is of little consequence.

A different set of conditions exist on the *valley rafter* that requires the use of *backing* or a *drop cut*. Figure 8-15a shows the frontal view of the lower end and top surface of an *unbacked*

TABLE 8-3   Backing or drop adjustment value
for nominal 2-in. hip or valley rafter

Rise (in.)	Backing or drop (in.)
2	1/8
3	5/32
4	3/16
5	7/32
6	1/4
7	9/32
8	5/16
9	11/32
10	3/8
12	7/16
15	1/2
18	9/16
24	5/8

Note: If a nominal 1-in. hip rafter is used, use one-half the value listed under the backing or drop column for the rise of the roof.

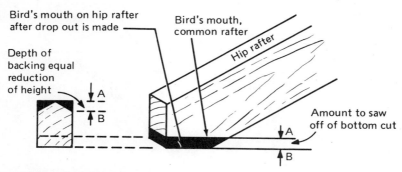

Figure 8-14   Drop cut on bird's mouth seat.

valley rafter. Notice that the level line is lower than the unbacked surface. Figure 8-15b is an illustration of a properly backed valley rafter. Notice that the lowest point of the V groove is in line with the level line.

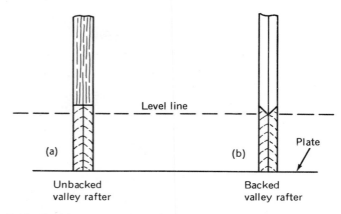

Figure 8-15 Backing on a valley rafter.

If a valley rafter is used without a backing or drop cut several conditions surface that create problems, as seen in Figures 8-16 and 8-17. In Figure 8-16 sheathing installed on the roof and intersecting in the valley *does not* bear on the valley. Notice that if sheathing is square cut (not beveled) its lower lip barely touches the valley. In many cases no contact is made. This condition is unsatisfactory. In Figure 8-17, the second condition, plywood or other sheathing is caused to lift away from the end of the jack rafter even though the sheathing is beveled. This allows sheathing full contact on the valley which is good; however, its lifting away from the jack rafter surface is not. In addition, this lifting causes some problems when cornices are constructed.

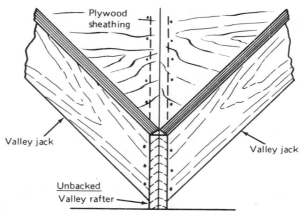

Figure 8-16 Unbacked valley rafter, sheathing does not make contact.

Figure 8-17 Unbacked valley rafter, sheathing mitered but buckled.

If the problems cited above are to be avoided, backing as in Figure 8-18 is one solution. Table 8-3 can be used to obtain the amount of backing (depth of V groove). Once it is defined and laid out on the valley rafter, a power saw is used to make the two necessary cuts. Look at Figure 8-18 and notice that a continuous line extends along the top lip of the valley jack rafter to the lowest point of the backing cut. Also notice that beveled sheathing fits snugly into the V groove and flush with the valley jack.

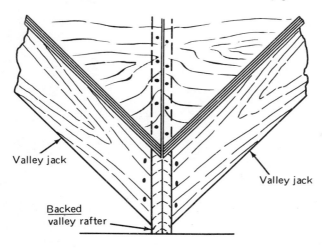

Figure 8-18 Backed valley rafter, properly seated and cut sheathing.

If a *drop cut* is used in lieu of the backing cut as in Figure 8-19, several adjustments must be made. One adjustment is made to the valley rafter, and another to each jack rafter. In Figure 8-19 an amount of drop is chosen from Table 8-3 according to the pitch (rise) of the roof and is shown as the blackened area (*G* to *H*).

Figure 8-19   Dropped valley rafter and jack rafter alignment.

Since the valley is lowered by the drop cut, valley jack rafters must be installed properly; that is, their top edges must align with the center line of the valley rafter. Figure 8-20 shows this idea more clearly. Each jack rafter is raised above the top surface of the valley so that the straightedge that lies along its surface touches the valley rafter as shown.

Figure 8-20   Aligning jacks on a dropped valley rafter.

Backing or a drop cut is a very important detail that must be included in the planning and cutting of either a hip or valley rafter, as is the adjustment that must be made during planning and cutting for rafter overhang.

Overhang

Figure 8-21   Hip and valley rafter tail (overhang).

The overhang of the hip and valley rafter performs the same function as on a common rafter; it may also be called *rafter projection* and *rafter tail.* However, this is where the similarity to the common rafter ends. Figure 8-21 provides the reference of the overhang necessary for this study. In Figure 8-21a the hip rafter is shown extending beyond the wall line, and in Figure 8-21b the valley rafter is shown extending beyond the wall plate also. While observing these illustrations, notice that the end cuts on the overhang are just like the ones on the flush rafter studied earlier. Because of the slightly different approach needed to explain and understand the overhang on each type of rafter, the hip is studied first, followed by the valley rafter.

When the overhang is used on a residential house it is usually made to project horizontally from the wall. The three most common dimensions of projection are the 6 in., 12 in., and 24 in. Figure 8-22a shows these as *run of common rafter projection.* The heavy black line represents the building line. It thus follows that on a hip roof the overhang extends on all walls as the illustration shows.

Figure 8-22   Hip rafter projected length.

On an equal-pitch hip roof the horizontal run of the hip roof is calculated as the hypotenuse of the right triangle whose sides are either 6 in., 12 in., or 24 in. With the figure on hand, the next step is to calculate the hip rafter *projected length*. For this the rise is used as the leg opposite the right angle. If these values are inserted into the pythagorean formula, the projected length is defined.

A simple on-site method is usually used by the carpenter. He simply steps off the framing square for the projection needed. However, a simpler method is available. All of the calculations necessary for any of the three projections are found in Tables 8-4, 8-5, and 8-6. To illustrate their ease of use and effectiveness consider the following three examples.

Example 1:   Determine the projected *hip rafter length* on a roof with a 5-in. pitch and planned 6-in. overhang.

1. Figure 8-22b indicates that the projected hip rafter length shall extend from the corner of the plate/wall line a distance of X.
2. From Table 8-4, for a 5-in. rise, the length is 9-13/16 in.
3. The hip rafter must be extended a length of 9-13/16 in. beyond the corner or original line of measurement.

Example 2:   A building is to have a 12-in. overhang. It has a rise of 8 in 12 on an equal pitch hip roof. Determine the *hip rafter projected length*.

1. Again, referring to Figure 8-22b the projected length of the hip rafter is defined as X.
2. From Table 8-5, for an 8-in. rise the length is 18-3/4 in.
3. The hip rafter projected length for a 12-in. overhang is 18-3/4 in. long.

Example 3:   An overhang of 24 in. is called for on the blueprints. The roof design is an equal-pitch hip roof whose pitch is 6 in 12. Find the hip rafter projected length (X).

1. For the third time, Figure 8-22b indicates the projection needed is X.
2. From Table 8-6, for a 6-in. rise the length is 34-7/16 in.
3. This indicates that the hip rafter projected length is 34-7/16 in. long.

TABLE 8-4.   6-in. Overhang
Hip and valley rafter overhang on an equal-pitch roof design.

Rise (in.)	Length (in.)
2	8-11/16
3	9
4	9-3/8
5	9-13/16
6	10-3/8
7	11
8	11-5/8
9	12-3/8
10	13
12	14-5/8
15	17-1/4
18	19-7/8
24	25-7/16

Note: Lengths have been rounded to the nearest one-sixteenth inch.

TABLE 8-5.   12-in. Overhang
Hip and valley rafter overhang on an equal-pitch roof design.

Rise (in.)	Length (in.)
2	17-1/16
3	17-1/4
4	17-7/16
5	17-11/16
6	18
7	18-3/8
8	18-3/4
9	19-3/16
10	19-11/16
12	20-3/4
15	22-5/8
18	24-3/4
24	29-3/8

Note: Lengths have been rounded to the nearest one-sixteenth inch.

TABLE 8-6.   24-in. overhang
Hip and valley rafter overhang on an equal-pitch roof design.

Rise (in.)	Length (in.)
2	34
3	34-1/16
4	34-3/16
5	34-5/16
6	34-7/16
7	34-11/16
8	34-7/8
9	35-1/4
10	35-3/8
12	36
15	37-1/8
18	38-7/16
24	41-9/16

Note: Lengths have been rounded to the nearest one-sixteenth inch.

Each of the tables provides at a glance the required hip rafter projected length for all of the most common pitches of residential roofs.

The materials estimating and planning phases for a hip rafter (and valley rafters) should be done in the office. In earlier paragraphs, explanation of how to obtain the basic hip or valley rafter lengths were given. It then becomes simple to refer to Tables 8-4, 8-5, or 8-6 to find the value of the projected length and add this length to the previous findings to determine full rafter length. It is even possible to use adjusted values from these tables for estimating overhang lengths where, for example, the run of common rafter projection is to be 8 in.

Finding the *valley rafter projected* length is no more difficult than finding the hip rafter projected length when the run of common rafter projection or overhang is 6 in., 12 in., or 24 in. Tables 8-4, 8-5, and 8-6 may be used. Figure 8-23a provides details that illustrate the run of valley rafter projection on a horizontal plane, which is the hypotenuse of the right triangle whose sides are either 6 in., 12 in., or 24 in. The value of the hypotenuse then becomes the base leg, and the rise is the opposite leg of a triangle that defines the length Y in Figure 8-23b. However, the value of Y can be selected from one of the three tables mentioned earlier (see examples for finding X and use the same approach for finding Y).

Bird's Mouth Adjustment When Hip and Valley Rafter Have a Tail

Hip and valley rafters require different bird's mouth cuts because of the tail projection. Figure 8-24 illustrates the require-

Figure 8-23   Valley rafter projected length.

ments where a miter cut is needed in each bird's mouth. Consider the hip rafter first where a corner miter is made to fit around the plates of the wall. The deepest point of the miter (corner) is the *original rafter length*. The tail length X projects from this point to the tail end. At the tail end angle miters are made as explained earlier.

Figure 8-24 Bird's mouth miters and tail cuts for hip and valley rafters.

If mitering is not used in the bird's mouth as shown then a square cut at the corner must be made. Quite frequently this miter cut is omitted because of the time it takes to cut it (mostly by hand with hammer and chisel). This attitude is prevalent where boxed cornices are employed. However, if open beam construction is employed the miters must be cut.

Contrast the bird's mouth cut in the valley rafter with the hip and observe that a miter is made whose angles are easily cut with a hand saw. No other option is available here; they must be cut so that the rafter seats properly into the corner of the wall and tightly against the plates. The end of the tail must have the internal miter as shown.

SELECTING THE PROPER SIZE AND GRADE OF RAFTER

The hip rafter must support live and dead loads as well as resist wind forces. It is a beam whose length is customarily long and all hip jack rafters must be securely nailed to it. Therefore, the selection of one by species and grade to accept and contain the loads distributed over its length must meet strength requirements as well as stiffness requirements of $S/180$, or $S/240$.

Additional strength and stiffness is automatically incorporated into a hip or valley rafter if the one used is 2 in. wider than the common rafter. This procedure is often elected to ensure that the side-bevel cut of the jack or cripple rafters have full nailing and contact surface. This means that:

- If common rafters are 2 x 6s, use a 2 x 8 for hip or valley rafter.
- If common rafters are 2 x 8s, use a 2 x 10 for hip or valley rafter.
- If common rafters are 2 x 10s, use a 2 x 12 for hip or valley rafter.

On occasion the hip or valley rafter is so long that a splice must be made. When this is done the splice must fall approximately mid-way between jack rafter on-center spacings as Figure 8-25 shows. A short block of rafter stock or plywood must be nailed securely to each side of the splice. When such an arrangement is used on a hip rafter there seldom is any problem since the rafter is in compression. However, if such an arrangement is used on a valley rafter additional support in the form of a truss brace should be installed from the rafter to a bearing partition. The valley rafter is in tension, and it supports the loads of the jack rafters, sheathing, and live loads. These forces are considerably more than the hip must support.

Figure 8-25 Splicing a hip or valley rafter with blocks of rafter stock.

Summary

Many different approaches to laying out and fitting hip and valley rafters are available from numerous sources besides this chapter. However, most of the other presentations employ only the framing square and the practical on-site method used in hip and valley layout. This chapter explains the math that is needed to understand why and how a hip and valley are laid out. It provides tables that at a glance provide values for hip and valley rafters on run, sidecuts, backing, and drop cuts and for the three most commonly used overhangs. Remember that this chapter deals *only* with the *equal-pitch* hip or valley rafter design.

PROBLEMS AND QUESTIONS

1. What is an equal-pitch roof design in a hip or valley?
2. How many hips are shown in Figure 8-1?
3. What type of rafter is used in dormer framing?
   a. hip   b. valley

4. In addition to a plumb cut at the ridge a hip or valley rafter probably uses a _____ cut.
   a. end   b. side
5. In an equal-pitch roof the hip or valley rafter is always at an angle of _____degrees from the exterior wall plates.
   a. 30   b. 45   c. 60   d. 90
6. Hip or valley run per foot run of common rafter is?
   a. 12 in.   b. 17 in.
7. If a common rafter run is 10 ft. and the rise is 4 in 12 what is:
   a. The run length of a hip rafter?
   b. Length of the hip rafter?
8. Write a short paper explaining why a bevel square is set for angles less than 45 degrees when marking side-cut bevels on hip or valley rafters.
9. From your study of Figure 8-8, at what angle would you set a portable circular saw table to make the side cut *if* the plumb cut line were already drawn?
   a. Cannot determine
   b. An angle of 45 degrees
   c. An angle of 90 degrees
10. Why must an adjustment for length be made to the ridge end of a hip rafter?
11. Explain why backing cuts need to be made to the top edge of each hip rafter.
12. What is dropping? How is the distance of dropping determined?
13. What would the estimated rafter tail length be for a hip that had a 5 in 12 pitch and planned projection of 12 in.?
    a. 12-in.   b. 18-in.   c. 24-in.
14. What size hip or valley rafter would you use if the common rafters were 2 x 8s?
    a. 2 x 8   b. 2 x 10   c. 2 x 12

9

The Roof, Part III:
Jack and Cripple Rafters

In each roof constructed with hip and valley rafters there is a need for jack rafters and sometimes cripple rafters. These are specially cut rafters that are actually modified common rafters. This chapter provides explanations and illustrations about both types of rafter. Definitions are given and comparisons are made between the hip, jack, and common rafters and the jack and cripple rafters. Specialized data are provided for the determination of rafter length and the different cuts.

DEFINING THE LOCATION OF A HIP, VALLEY, AND CRIPPLE JACK RAFTER

Figures 9-1a and b illustrate two roofs that include the rafters we will study in this chapter. Both roofs are defined as hip and valley roofs. In Figure 9-1a all ridges are at the same height, whereas in Figure 9-1b one ridge is much lower than the other. Even so, the principles of jack rafter identification remain the same.

On the left of Figure 9-1a *hip jack* rafters extend from the building plates to the hip rafter. This illustration shows six hip jacks, four on the end and two on the side. Notice that the upper end of each hip jack is connected to the hip rafter.

The right of Figure 9-1a illustrates a valley and *valley jacks*. These are rafters that are connected to the ridge and extend downward to the valley rafters where their lower ends connect. The illustration shows eight valley jacks. Figure 9-1a does not show any cripple jacks.

(a)

(b)

Figure 9-1 Framed hip and valley roof design using jack and cripple rafters.

Figure 9-1b illustrates a roof containing all three types of jack rafters. On the left, the lower roof uses four valley jacks, and the higher roof uses three valley jacks. The lower roof does not use any hip jacks but the higher one uses five hip jacks. Between the valley and hip on the higher roof are several rafters that do not touch either the wall plate or ridge. These rafters are defined as cripple jacks. Notice that there are three between valley and hip. Also notice that there is one other cripple jack connected between both valley rafters (left of longer valley rafter). It is classified as a cripple jack because it does not connect to either the plate or ridge.

These definitions of jack rafters are always true regardless of the roof design. But, even though the hip jacks rest on plates and valley jacks are connected to the ridge, could there be correlation between these rafters and the common rafter to make understanding them easier?

DETERMINE THE COMMON AREAS OF JACK AND CRIPPLE RAFTERS WITH COMMON RAFTERS

Rise and Run Values

The rise and run values of a hip jack and valley jack are the same as those of the common rafter in the same roof. This means that if, for example, a roof has a 4 in 12 pitch the hip and valley jacks use the 4 in 12 too. How and where are the rise and run or pitch used? These values are used for laying out the bird's mouth and rafter tail cuts on the hip jack, and to lay out the ridge cut on the valley jack.

Figure 9-2 shows a side view of a hip jack as though looking through the hip so that the correlation of rise and run are understood. Notice that all the hip jacks are in alignment with common rafters. This is as it should be. If a similar view were prepared for the valley jack, their alignment with the common rafter would agree. However, the common connecting point would be at the ridge.

Bird's Mouth and Rafter Tail

Figure 9-2 shows that the bird's mouth and rafter tail of each hip jack rafter is identical to the common rafter's bird's mouth and rafter tail. The pattern that was designed to lay out this portion of the common rafter should be used when laying out the hip jack rafter. Since the length of the jack will be shorter than the common rafter, it is a good idea to draw a plumb line to the rafter's top edge as the figure shows. The rafter's length can then be measured along the top line.

Figure 9-2 Comparing the hip jack rafters to the common rafter.

Ridge Cut

Each valley jack rafter needs to have a common rafter ridge cut. This cut is made using the roof pitch and framing square layout, followed by cutting. Normally one-half the ridge board thickness is subtracted from the common rafter's length before the ridge cut is completed. This allowance is necessary when valley jacks are cut. However, rather than cutting it from the ridge cut, the allowance can be made a part of the compound miter cut at the lower end of the rafter. Figure 9-3 shows four valley jack

rafters spaced evenly as they appear with the valley rafter removed. Carefully note that the lower end of each jack has a compound miter cut. One part of the cut is the pitch; the other part is
the bevel needed for the rafter to join the valley rafter.

Figure 9-3 The valley jack rafter spaced properly, fastened to ridge, and
revealing the compound miter cut needed to seat against the valley rafter.

Cripple Rafter and Jack Rafter

The cripple jack rafter does not touch either the plate or
ridge, yet its pitch is the same as the common rafter and jack
rafter. It therefore must fall on a plane with these other rafters as
Figure 9-4 shows. Since it touches neither plate nor ridge, two
compound miter cuts are needed. One cut is made so that the
cripple seats properly against the hip rafter, the other cut is made
so it seats against the valley rafter. (View Figure 9-1b if reorienta-
tion of the cripple jack is needed.) It also may be that the compound miter is made to fit the rafter between two valley rafters. In
any event the plumb cut for the common roof pitch is one of the
two bevels used in the compound miter.

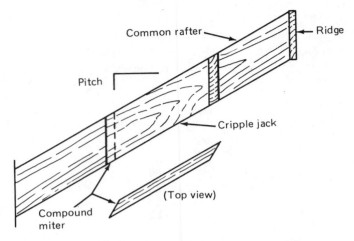

Figure 9-4 Comparing the cripple jack to the common rafter.

DETERMINING THE JACK RAFTER LENGTH AND MITER CUTS

There are several ways to determine the length of a jack rafter. One is by physically measuring its length on the roof assembly; another is by using the jack rafter tables on a framing square and a third is by calculating the length. The two most practical are the measuring techniques and using the rafter table on a framing square. Sometimes a combination of these techniques simplifies the task.

Practical Measuring Method

Figure 9-5 illustrates the practical measuring method of finding the length of the first hip jack when the roof layout of rafters is from right to left. The last installed common rafter is the reference from which all calculations begin. On-center spacing is marked at the top edge of the hip (in this example, 16 in.). The plate is already marked. However, a level-line (chalkline) should be established as shown and the rafter length is measured with a steel tape. Notice that the length obtained is from the *longest* point of the compound miter and does *not* include any rafter tail.

Figure 9-6 shows that this method can be used for finding the length of the shortest hip jack. In this situation the layout of on-center spacing is considered to be left to right. Once again the jack length is from its longest point to the level line.

Figure 9-5 Practical method of defining the longest hip jack rafter.

Figure 9-6 Practical method of defining the length of hip jack rafter, shortest rafter first.

A handy aid that is easily made on site is the *marking* stick, shown in Figure 9-6. A 1 x 2 or 1 x 4 laid off at on-center spacings from plate markings can easily be tacked to common rafters and hip rafter. A word of caution; the marking stick must be parallel to the level line for accurate jack lengths to be determined.

The methods used in finding the hip jack length are easily adapted to finding the valley jack lengths. Figure 9-7 shows that the longest valley jack's length is found with square and tape measure. The square provides a true on-center point on the valley rafter. The measure of the valley jack is from this point to the ridge. Notice that the point is the tip of the bevel; however, this point is not the longest part of the rafter layout since the pitch cut must be made from this point as well. This idea is only brought out to preclude one from assuming incorrectly that a rafter of roughly this length is sufficiently long enough. Allowance must be made for the plumb cut.

The laid off marking stick may also be used as an aid in determining the valley jack rafter length. If used, it must be a-ligned parallel with the ridge rather than the plate. One final word about this method; it is equally well suited to determining hip or valley jack rafter lengths for on-center spacings of 24-in. or other on-center spacings.

Figure 9-7 Practical method of laying out on-center spacing and defining length of longest valley jack rafter.

Framing Square Table Method

Many of the more expensive framing squares provide tabular data on the length and difference in length of jack rafters. They are frequently titled "Difference in length of jack rafters" because the length of the first shortest or first longest hip or valley jack is variable. However, once the length of either is known the table data is appropriate to use.

Table 9-1, which is a part of Figure 9-8, lists most of the common roof pitches as well as one or two extremes. Notice that this table is for the difference in length of jack rafters spaced 16-in. o.c. as might be found on a framing square. Also notice the twelve right triangles below the table. These are used to help explain the table and orient the carpenter.

Assume a roof has a 3 in 12 pitch. The difference in length of the jack rafter in Table 9-1 is shown as 16.49 in. The triangle with the 3 included shows how this value compares with (1) the 16-in. o.c. spacing and (2) the hip or valley length/ft. run. A translation to meaning is that for each 16 in. of spacing along the plate or ridge line the jack rafter's length must be increased 16.49 in. or 16-1/2 in. The length of the jack rafter from the point of view of the hip or valley must be increased or decreased in length 22.97 -in.

Table 9-2 provides quick reference of the same kind of data but for 24-in. o.c. spacing of jack rafters. For a 3-in. pitch where jacks are spaced 24-in. o.c. the difference in length is 24.74 in. or 24-3/4 in.

The table and right triangles shown in Figure 9-8 do not always lend themselves to translation that is meaningful. Therefore, four examples have been chosen that should make their meaning clear.

Example 1: Figures 9-9a and b illustrate two *hip roof jack rafter* situations that make use of the tabulated data. In Figure 9-9a a pitch of 3 in 12 is shown. An orientation of the figure is necessary before explanation can proceed. Locate the following points:

- Hip rafter line. This is the line that the hip rafter follows from plate to ridge.
- Wall plate line. This is the horizontal wall plate that the jack must seat upon.

Table 9-1 Difference in length of jack rafters (16 in o.c.)

Rise	3"	4"	5"	6"	7"	8"
Diff/16" o.c.	16.49	16.87	17.33	17.875	18.52	19.23

Rise	9"	10"	12"	15"	18"	24"
Diff/16" o.c.	20	20.83	22.625	25.61	28.84	30.72

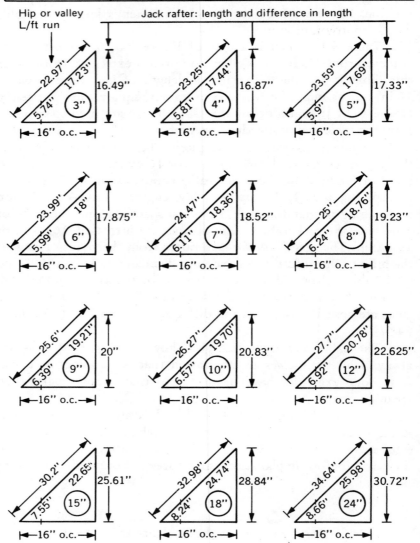

Figure 9-8 Mathematically determining the length and difference in length of jack rafters for numerous pitches of roof.

TABLE 9-2. Difference in Length of Jack Rafters (24-in. o.c.)

Rise	3 in.	4 in.	5 in.	6 in.	7 in.	8 in.
Diff/24-in. o.c.	24.74	25.30	26	26.83	27.78	28.84
Rise	9 in.	10 in.	12 in.	15 in.	18 in.	24 in.
Diff/24-in. o.c.	30	31.24	33.94	38.42	43.27	

- 16-in. o.c. The spacing of the jack rafters.
- 3 hip jack rafters.
- 16.49 in. The difference in length of the jack rafters taken from Table 9-1.
- 16.49 in. Converted to 16-1/2 in.
- 44.1 degrees. This is the angle of one miter of the compound miter.

The *dashed line* that is parallel with the *wall plate line* begins at the intersect point of the shortest jack and bisects both jacks to the right. From this line the difference in length of jacks is determined from Table 9-1. The length needed to be added to the second jack is 16.49 in. or 16-1/2 in. Therefore, its total length is x + 19-1/2 in., where x is the length of the first jack. The third jack has a length of x + 19-1/2 in. + 19-1/2 in. because a step of 16 in. to the right calls for an additional 16-1/2 in. of jack rafter length.

If a ruler were used to measure the hip rafter line between any two similar points where jacks intersect the hip, its length would be 22.97 in. or almost 23 inches. One other important factor in the drawing is the 44.1 degree miter needed at the tip end of the hip jack. Each pitch results in a different angle. Zero pitch, for example, would result in a 45 degree miter. (A table with these miters is given later in this section.)

Example 2: The significant difference in Figure 9-9b is that all data are for an 8 in 12 pitch. The identities of hip rafter line, wall plate line, three jack rafters, and 16-in. o.c. spacings are the same as in Example 1. However, a much steeper roof is being constructed when the pitch is 8 in 12. Therefore, the difference in length of the jack is predictably longer. Table 9-1 calls out a difference in length of 19.23 in. The second jack shows that its total length is x + 19.23 in. or x + 19-1/4 in. L.

Figure 9-9 Laying out the hip jack rafter for difference in length and defining the number of side cut degrees.

What is the L for? Since 0.23 in. is just short of 0.25 in., which is equal to 1/4 in., the 19-1/4-in. mark must be termed *light*. That is, the carpenter makes a mark on the lumber at a length of $x + 19$-1/4 in. He then cuts the stock so that almost all or all of the mark is cut away too. By this adjustment to the cutting, he accounts for the slight discrepancy that was introduced in converting hundredths to fractions. It also means that as he lays out the third jack he must make two light (L) adjustments. One he makes by using the very leading edge of his pencil mark, and the other by cutting away the line. Needless to say these are sophisticated adjustments that are used by the experienced carpenter and must be learned by the apprentice.

Example 3: Figures 9-10a and b provide data on *valley jack rafters* for the 10 in 12 and 5 in 12 pitch roof similar to that given in Figure 9-9. However there are several differences.

- Roof ridge line. A roof ridge line must be the horizontal reference for valley jacks instead of the plate line.
- Valley rafter line. This line is the position of the valley rafter.
- Difference in length. The difference in length may be considered either from shortest to longest or vice versa.

The *dashed lines* parallel with the *roof ridge line* (Figure 9-10) show the point of intersection of the valley rafter and jack rafter. The difference in length for this pitch is 20.83 in. So the length of the jack *at the shortest point* of the second jack is equal to $x + 20.83$ in. or $x + 20$-13/16 in. F. (F stands for full). No problem exists using this method since additional rafter stock needed for the compound miter is already understood to be necessary. Note also that the angle of the miter is 37.5 degrees which is quite a bit different than the 44.1 degrees used with the 3 in 12 pitch.

Example 4: Figure 9-10b shows a 5 in 12 pitch roof with valley and valley jacks. Data are similar to the other three examples except that the difference in length is 17.33 in. as Table 9-1 in Figure 9-8 lists. As before, the total length of the longest jack shown is $x + 17$-5/16 F + 17-5/16 F.

Full means that the thickness of each marking line must be used and it must still be on the stock after the cut is made. In the

Figure 9-10 Laying out the valley jack rafter for difference in length and defining the number of side cut degrees.

conversion of 0.33 to 5/16 in. there is a discrepancy: 5/16 in. = 0.315. The thickness of a pencil line will just about make up the 0.015 shyness.

Jack Rafter Compound Miter Cuts

With this orientation and determination of the jack rafter length the compound miter cut can be examined. Since the four examples just studied provide a variety of pitches as well as examples of hip and valley jacks, their characteristics are used in explaining the compound miter cut.

It should be understood that the compound miter cut of all jacks fit against either the hip rafter or valley rafter. It is also important to know that as the rise of a roof (pitch) increases the bevels of the compound miter become more acute. Figure 9-11 shows the four jack rafters previously illustrated in Figures 9-9 and 9-10. Notice the following data as support for the previous statement of pitch vs. miters:

Pitch	Degree
3 in 12	44.1
5 in 12	42.7
8 in 12	39.7
10 in 12	37.5

It is shown that the plumb cut angle increases with a rise in pitch and also that the side cut angle becomes a more acute angle. Obtaining the plumb cut layout with the framing square is relatively easy, but laying out the side cut can be difficult if degrees are used. There is a purpose for listing the degrees as shall be shown.

Three methods are available for use in laying out and cutting the side cut of the jacks.

1. The framing square provides one solution. Either the table imprinted on the square for side cuts should be followed or the one in Table 9-3 should be used. If Table 9-3 is used, column 3 provides the combination of values needed for the proper degree miter required. For example, to lay out the side cut for a 3 in 12

Figure 9-11 Laying out the compound miter cut for various pitch roof jack rafters.

pitch, place 12 in. (blade) and 12-7/16 in. (tongue) such that this last value intersects with the longest edge of the plumb cut; then mark alongside the tongue.

2. The second method uses a bevel or miter square and framing square as Figure 9-12 shows. A bevel square is set to a predetermined value and locked tightly with its wing nut. This tool is then used to lay out all side cuts for jacks that have common rises. (Refer to Table 9-3 and Figure 9-12 throughout the explanation).

TABLE 9-3.   Side Cuts for Jacks and Cripple Rafters

Rise	Degrees	Framing square 12 in. on blade tongue	Miter square on framing square[a]	
			$(x = y/Tan)$[b]	1-in.[c]
3	44.1	12-7/16 in.	2-1/16 in.	1-9/16 in.
4	43.5	12-5/8	2-1/8	1-5/8
5	42.7	13	2-3/16	1-11/16
6	41.8	13-7/16	2-1/4	1-3/4
7	40.8	13-15/16	2-5/16	1-13/16
8	39.7	14-7/16	2-7/16	1-15/16
9	38.6	15	2-1/2	2
10	37.5	15-5/8	2-5/8	2-1/8
12	35.3	17	2-13/16	2-5/16
15	31.9	19-1/4	3-3/16	2-11/16
18	29.0	21-5/8	3-5/8	3-1/8
24	20.5	32	5-3/8	4-7/8

[a]See technique shown on Figure 9-12.
[b]Formula assumes $y = 2$ in., the width of framing square blade; values rounded to the nearest one-sixteenth inch.
[c]Readings on inner blade scale; miter square's steel edge set on 1 in.

The blade of a framing square is 2 in. wide; the number of degrees for side cut per rise is known; therefore using the formula:

$$x = \frac{y}{Tan}$$

provides a conversion that defines the values listed in column 4 of Table 9-3. Note in a right triangle these values would be the base, and 2 in. would be the opposite side.

Since the *outer corner* of a well-used square receives much abuse, its exact position may not be exactly defined. Therefore, the setting of the bevel square should be made from the *1 in.* mark as shown in Figure 9-12. It is also standard that the framing square's tongue be 1-1/2 in. wide, so this value has been subtracted from the value in column 4 of Table 9-3 and the remainder is shown in column 5.

The blade on the miter square should be set according to column 5 of Table 9-3 and Figure 9-12 for each side cut bevel needed. For example, if an 8 in 12 pitch is used the bevel square is

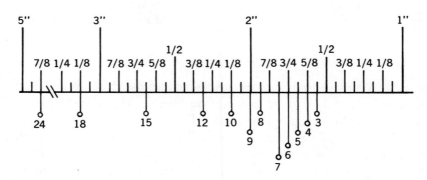

Figure 9-12   Bevel square set ups for jack and cripple rafter side cuts.

positioned as shown in Figure 9-12a, where the edge of the blade is set against the 1 in. on the outer numbers and 1-15/16 in. on the inner numbers. The scale at the bottom of Figure 9-12 is expanded to show the points that should be used with various pitches if the bevel square technique is used.

3. The third method uses the degrees listed in Table 9-3 and a radial arm power saw. This power saw's tilting head is set to the number of degrees shown in column 2 of Table 9-3 and its arm is set for the plumb cut. No marks for side cuts need to be made on jacks *after* a test cut has been verified as accurate. With the saw set properly a single operation cuts the compound miter.

LENGTH AND BEVEL CUTS ON A CRIPPLE JACK

Once the proper orientation of a cripple jack rafter is understood, the practical determination of its length and compound miter cuts become relatively easy. Refer again to Figure 9-4, which shows the orientation of the cripple jack to common rafter. This figure shows that the rise per foot run of the cripple jack is the same as the common rafter. This also means that the hip and valley jack and cripple have common elements. Their compound miters are the same. One of the bevels of the compound miter is the plumb cut and the other is the side cut. These are found using the method for hip or valley jacks from Table 9-3 and Figure 9-12.

Practical Method of Determining the Cripple Jack Length

The simplest and most practical method of finding the length of the cripple jack is shown in Figure 9-13. It involves the use of a 1 x 2 piece of stock, a ruler, and mason line.

Assume that cripples are spaced 16 in. o.c. Assume further that the roof being framed resembles the one shown in Figure 9-13 and that a hip jack is already in place. The object is to find the length of the first and second cripple jacks.

1. Mark a point on the valley rafter which is 16 in. o.c. to the right of the hip jack and partially drive a nail into the valley.
2. Install a 1 x 2 as shown, making sure it is parallel to the plate. Mark on the 1 x 2 a point 16 in. o.c. as shown.
3. Install a mason line over the nail and center it over the mark on the 1 x 2 and where the line contacts the hip rafter (*b*) make a pencil mark.
4. Remove the mason line and measure the distance between the points *a* and *b*. This gives the length (*L*) of the cripple.

Figure 9-13 Practical method for defining the length of the cripple jack rafter.

Notice that the difference in length can be used; however, the length from point *a* to the beginning of the difference in length point *c* needs to be measured. So why complicate the practical method?

Compound Miter Cuts on Cripple Jacks

With the length of the cripple defined, the compound miters need to be laid out and cut. The procedure is exactly the same as that used for hip and valley jack rafters.

Rather than elaborate the methods previously explained, suffice it to say that they may be used if the following suggestions are taken:

1. Use the same pitch as on the main common roof rafter to mark the plumb cuts.
2. Use the proper setting of bevel square or framing square.
3. Make sure the lines drawn on the cripple jack stock are properly oriented according to their position of installation.

Since both cripples in Figure 9-13 and in most actual situations are exactly the same, cut two or as many as needed. Remember that when nailing these members in place they must align with their on-center spacing and be properly oriented for height to backed or dropped hip and valley rafter.

PROBLEMS AND QUESTIONS

1. What cut must be made to a common rafter to make it a jack rafter?
2. A hip jack rafter employes a
   a. ridge cut   b. bird's mouth cut
3. A valley jack rafter uses a
   a. ridge cut   b. bird's mouth cut
4. Why does a cripple jack rafter have a compound miter cut on each end?
5. Are the rise and run values for a jack rafter the same as for *common* or *hip* rafters?
6. Describe in a few paragraphs the practical method of determining the shortest jack rafter length.
7. Assume a roof has a 6 in 12 pitch and rafters are spaced 24-in. o.c. The shortest jack is 23 in. long; what is the length of the next jack?
   a. 46 in.   b. 49-5/8 in.   c. 50-7/8 in.
8. Why is the jack rafter's side cut line an angle less than 45 degrees when measured along the crown of the rafter?
9. Lay out and mark a 5 in 12 pitch for a common ridge cut on a scrap 2 x 8. Set your portable circular saw to 45 degrees and cut along the line. Set your miter square so that its body is even with the top edge of the 2 x 8 and adjust and tighten the blade to the bevel cut made with the saw (not ridge cut). Next lay the miter square as shown in Figure 9-12 and compare the values. Did the cut produce a value equal to the last column in Table 9-3? (See note C in the table.)

10

The Roof, Part IV:
Other Requirements
and Sheathing Techniques

Many times a picture and a few carefully chosen words can develop more fully an understanding of a complex topic. If a serious study has been undertaken thus far, the selected items of this chapter will easily fall into place with the understanding previously gained.

To avoid tiresome repetition of adaptations of principles previously covered, only the aspects of principles and tasks related to other framing needs are given. Subjects included are dormer construction, collar ties, gable end studding of a gable roof, installation of a fly rafter, and framing for a chimney that passes through a roof. The remainder of the chapter identifies sheathing techniques used on roofs. Sheathing of proper thickness grade and installation are all important aspects of roofing that must be carefully attended to.

DORMER CONSTRUCTION

Dormer Without Sidewalls

The first in the series of dormers examined is the *dormer without sidewalls*. Figure 10-1 illustrates the framing technique used. Inclusion of such a dormer is rather expensive. Note the complexity of its construction. Initially a frame of headers and double rafters are constructed. Double headers are installed perpendicular to plate and ridge. The double common rafter, sometimes called *trimmer* rafter, is needed to supply additional strength. *Cripple common rafters*, sometimes called *jack common*

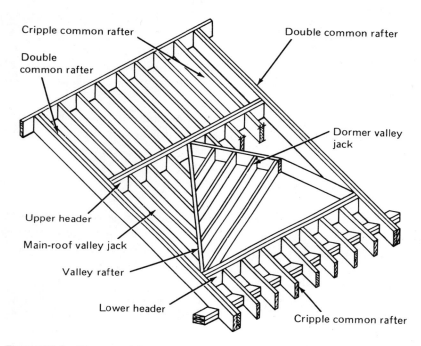

Cripple common rafter

Double common rafter

Double common rafter

Dormer valley jack

Upper header

Main-roof valley jack

Valley rafter

Lower header

Cripple common rafter

Figure 10-1 Dormer without sidewalls.

rafters, are installed with normal on-center spacing. Two valley rafters are installed next, followed by the installation of the main roof jacks and the dormer ridge and jacks.

From the details shown here, one new principle of roof construction is identified. It is the framing of the opening with headers and double common rafters. Note that the headers are cut from rafter stock. Their flat surfaces are not installed plumb as is the ridge; rather they are nailed so that both top and bottom edges are flush with the rafter.

What significance does this header installation have on valley rafters and dormer ridge board layout, cutting, and installation? For one thing, each valley rafter has a single side miter cut rather than compound miter. The top cut is probably a 45 degree bevel, the lower end of the valley rafter uses two 45 degree cuts to fit into the corner. Neither end uses a plumb cut, which is different from normal valley rafters. Installation is the same as explained in Chapter 8; no special problems should be encountered.

Main roof jacks will only need single miter cuts since they, like the valley rafter, are on a flat plain. The opposite (upper) ends use square cuts. In contrast, however, standard procedures need to be used for laying out and cutting the dormer jack rafters since these elevate to a different ridge. Chapter 9's methods should be employed here.

Cripple common rafters need to be laid out from the pattern for common rafters and then measured for length, cut, and installed.

Dormers with Side Walls

A dormer with side walls may be constructed where the walls rest on double rafters or where they are built up from studs resting on the ceiling joists. Both types are shown in Figure 10-2. In either case the studs and plates are usually constructed after the full roof joists are installed.

In the dormer shown in Figure 10-2a, the studs are cut and installed to rest on doubled rafters and headers. Two valley rafters are used and they connect to the header near the roof's peak and rest on the double rafters. Four jack rafters are needed as well as several short common rafters for the dormer roof.

Once again specialized layouts and cuts need to be made on the valley rafters where they are nailed to the header and dormer ridge. Then, too, the lower end of each valley intersects with dormer plate line and common rafter, so a special layout is needed here as well.

The cripple common rafters below the double header will need to be *plumb cut* since it is important that the header be installed plumb.

In Figure 10-2b a standard wall-framing situation exists that simplifies the roof construction somewhat. For one thing, a single lower header is all that is needed to support the rafters. Recall though that if the header length exceeds 4 ft double headers should be used.

In framing the roof two short valleys are needed as well as four jack rafters. Notice that the lower end of the dormer jacks need a compound miter cut similar to those studied in Chapter 9. Both situations shown in Figure 10-2 will require individualized fitting of valley and jack rafters.

Figure 10-2 Dormers with sidewalls extending to the second floor or fastened to doubled rafters.

231

Shed Dormer with Side Walls

Figure 10-3 shows a typical shed dormer. The dormer opening must be framed in first. Double headers are installed plumb, which means common cripple rafters need plumb cuts where they contact the headers. Once the headers and common rafters are in place, the side walls are framed. Each shed roof rafter must be cut to fit against a previously installed common rafter. Obtaining the angle cut needed for a good fit is easily accomplished with a bevel square and uncut rafter or scrap 2 x 4 and rafter. After cutting a sample rafter and assuring it fits well, all others should be cut from this pattern.

Dormer rafter

Header

Common cripple rafter

Figure 10-3 Dormer with a shed roof design.

GABLE END STUD INSTALLATION

Once the rafters have been installed, the gable ends on the gable roof need to be cut and installed. There is more than one way to perform this task. However, one principle should always be used regardless of the method selected. That is the gable end studs should align over the wall studs which are in the end walls.

Figures 10-4 and 10-5 illustrate the two techniques for installing gable end studs. In Figure 10-4 each gable end stud is measured and cut for length. Then a cut out is made along its face edge
so that its outer edge aligns with rafter and wall plate. Toenailing
must be used on the lower end of each stud, but 12d or 16d nails
may be used on the upper end since these may be driven straight
through.

Figure 10-4   Installing gable end studs.

In Figure 10-5b, a special 2 x 4 plate is installed so that its
top surface is exactly positioned as a seat for the lookouts. Gable
end studs are cut with a bevel and the assembly is made accordingly. Once again, notice that each gable end stud is in alignment
with the wall studs below.

FLY RAFTER INSTALLATION

A *fly rafter* is used on a gable roof where the gable end needs an
overhang. Two methods are employed in Figure 10-5. In Figure
10-5a, standard common rafters are installed, and gable end studs
are attached as previously explained. A method of layout for *lookouts* is selected such as 16-in. or 24-in. o.c. spacing from lower

x = overhang less 1½"
   for fly rafter
y = total length of
   lookout

Lookouts

Lookout
layout
16" or 24" o.c.

Notches for
lookouts

90°

Y

X

Ridge cut

Fly rafter

(a)

Ridge

Rafter

Hanger

Double
rafter

Lookout

Plate

Top plates

Nailing block

Fly rafter

Gable end studs

End wall studs

(b)

Figure 10-5 Fly rafter installation techniques.

234

edge to ridge or vice versa. Notch marks are laid out and cut as shown so that each lookout sits neatly in place.

When all lookouts are installed, the fly rafter, which is usually a 2 x 4 or 2 x 6, is cut and then nailed to each lookout. Cornice is then applied over this frame.

The second technique is shown in Figure 10-5b. In using this technique, the last rafter before reaching the end wall is doubled, metal hangers are installed at 16-in. or 24-in. o.c. spacings as shown. The gable ends are completed. The nailing block shown aids in stabilizing the lookouts.

The fly rafter is cut to fit, then nailed against the lookouts. The fly rafter is usually as wide as the lookouts but may on occasion be wider.

FRAMING FOR CHIMNEY AND CHIMNEY SADDLE

Whenever a chimney is built to pass through the roof a *chimney well* must be framed. This well is similar to a stairwell in a floor assembly, but in most cases much smaller. Figure 10-6 illustrates a completed chimney well. This example shows that one common rafter is cut to allow the construction of the chimney well. It also indicates that doubled rafters are used on either side of the opening and headers complete the opening.

Figure 10-6b illustrates a cut-away view of the opening to dramatize the two basic standards that several building codes specify: they are the 2-in. clearance between framing and chimney and the 3/4-in. clearance between sheathing and chimney. These two requirements should be followed faithfully since a fire hazard is greatly reduced. Another reason for following this rule is to allow for roof movement, expansion and contraction as weather and wind exert their forces on the roof. The chimney is rather stiff by comparison and the spacing removes the possibility of rupture of the chimney.

The final part of roof framing around a chimney well is the construction of a *chimney saddle*. Figure 10-7 shows the saddle as a built-up area on the back side of the chimney. The use of the saddle causes the normal water runoff on the roof to shed to either side of the chimney rather than buffet it.

The peak of the saddle is level. The supports for it are made from nominal 2-in. stock tapered to form its shape. A 2 x 4 is

Figure 10-6   Framing for chimney.

usually used as the horizontal member and backing of its top edge is done on the best jobs. Sheathing is cut and nailed over the frame. Composition shingles similar to those used on the main roof should not be used on the saddle because of the chance that water will flow back under them and cause roof leaks. A special sheet metal cover should be made and installed. This cover should be made so that (1) it runs under several courses of shingles,

Figure 10-7 Chimney saddle framing and sheathing.

(2) covers the saddle, and (3) bends up the side of the chimney approximately 4 to 6 in. Chimney flashing will be installed over the saddle metal cover and leaks will not occur.

PLYWOOD SHEATHING

Plywood is customarily installed on the roof as sheathing. It is easy to install, covers the roof quickly, and various thicknesses may be used depending upon loads and rafter on-center spacings.

The thickness to use is easily determined from tables such as Table 10-1, which was prepared by the American Plywood Association.

The first column, titled "Identification Index," lists the rafter on-center separations left of the slash beginning with 12 in. and ending with 48 in. Numbers to the right of the slash are floor or ceiling joist on-center separations. The 2-4-1 and 1-1/8 in.

TABLE 10-1 Plywood Roof Decking

Identification Index	Plywood thickness (inches)	Maximum span (inches)	Unsupported edge-max. length (inches)	Allowable live loads (psf) — Spacing of supports center to center (inches)									
				12	16	20	24	30	32	36	42	48	60
12/0	5/16	12	12	150									
16/0	5/16, 3/8	16	16	160	75								
20/0	5/16, 3/8	20	20	190	105	65							
24/0	3/8	24	20	250	140	95	50						
	1/2		24	385	215	150	95						
32/16	1/2, 5/8	32	28		330	230	145	50	40				
42/20	5/8, 3/4, 7/8	42	32			300	190	90	75	50	35		
48/24	3/4, 7/8	48	36				225	120	105	65	45	35	
48/24*	3/4, 7/8	48	36					125	105	75	55	40	
2·1	1-1/8	72	48				390	245	215	135	100	75	45
1-1/8" Grp. 1 & 2	1-1/8	72	48				305	195	170	105	75	55	35
1-1/4" Grp. 3 & 4	1-1/4	72	48				355	225	195	125	90	65	40

*Loads apply only to C-C EXT-APA, STRUCTURAL I C-D, INT-APA, and STRUCTURAL I C-C EXT-APA. Check availability before specifying.

238

plywood designations are especially thick plywood panels most appropriately used in plank and beam construction. In a roof designed as an open cathedral or exposed rafter-beam type these last three types of plywood could be used as sheathing provided the spans between rafter beams did not exceed 72 in. (column 3).

If, for example, a roof is constructed with rafters spaced 16 in. o.c., either 5/16 or 3/8-in. thick plywood panels may be used to sheath it. Under the live load column, notice that a 75 psf live load maximum is indicated. To assure that this load will be borne by the plywood attached to the rafters, only the grades listed below the chart should be used.

If deflection is a concern, the plywood thicknesses and grades are applicable to $S/180$ for combined live and dead loads or $S/240$ for only live loads.

Leave 1/16" space at all panel end joints and 1/8" space at all panel edge joints. When wet or humid conditions can be expected, double these spacings.

Plywood sheathing

Plyclips or tongue & grooved edges if required

Detail

plyclips

Asphalt, asbestos, or wood shingles. Follow roofing mfr's. recommendations for roofing felt.

Protect edges of Interior plywood against exposure to weather, or use Exterior plywood starter strip

Exterior plywood soffit

Figure 10-8 Details for sheathing a roof.

Figures 10-8 and 10-9 illustrate several parts of the techniques of installing plywood on the roof. In Figure 10-8 ply-clips are shown being used between rafters to aid in supporting the edges of the plywood. These clips are frequently used with 5/16-in., 3/8-in. and 1/2-in. plywoods, where rafters span 16 in. or 24 in. o.c. They are not as a rule used where 3/4-in. plywood sheathing is installed over rafters with these on-center spacings. However, should rafters be spaced 32-in., 42-in., or 48-in. o.c., ply-clips should be used with the thicker plywood panels.

Figure 10-9   Sheathing a roof with plywood that includes a valley.

Figure 10-9 shows two installation techniques. One is the sheathing of the valley. Pieces of plywood are easily laid out and cut to fit as *Callout A* shows. These panels are best cut on the ground for safety sake. The other technique is trimming excess plywood along the rake of the roof. Normally a chalkline is snapped then a power saw is used to trim the plywood.

Nailing should follow the rule that the portion of nail penetrating the rafter should be at least one-half the nail's length. From this rule the following table of nail sizes versus plywood thickness is established.

TABLE 10-2 Nails for Sheathing	
Nail size	Plywood thickness (in.)
6d	5/16, 3/8, 1/2
8d	1/2, 5/8, 3/4
10d	3/4, 1 1/8
12d	1 1/8

Nailing should follow the schedule of 6 in. along the ends of the sheet and not more than 8-in. separation on intermediate rafters. Closer spacing of nails than just indicated adds strength to the roof and aids in its ability to withstand forces placed upon it.

Gluing panels to rafters has not been customary in the past. However, there is considerable merit to the idea of using glue. A much more uniformly sound structure is created. The bonding accomplished with the glue does make a panelized structure of considerable strength.

PROBLEMS AND QUESTIONS

1. What is the basic difference in construction between a dormer without sidewalls and one with sidewalls?
2. Why is a set of headers and doubled rafters installed first when framing a dormer without sidewalls?
3. What name is given to the doubling (second) rafter?
   a. jack   b. trimmer
4. Explain where cripple common rafters are used around the area of the dormer.
5. How many valley rafters does each dormer framework with side walls require?
   a. 2   b. 4   c. 6
6. In a dormer with sidewalls do the studs always extend to the floor joists of the second floor?
7. Where should the upper end of each shed-type dormer rafter end?
8. What is the principle of gable end stud alignment?
9. Which type of roof may use a *fly rafter*?
   a. hip   b. gable

10. What is the purpose of a lookout in fly rafter installation?
11. What is the standard acceptable clearance between chimney and framing for residential houses?
    a. 1 in.   b. 2 in.   c. 3 in.
12. What is the recommended clearance between sheathing and chimney?
    a. 1/2 in.   b. 3/4 in.   c. 1 in.
13. Describe the purpose of a chimney saddle.
14. What would be the *minimum* thickness of plywood sheathing for use on a roof with rafters spaced 16-in. o.c., where a live load of 75 psf or less is predicted?
    a. 1/4 in.   b. 5/16 in.   c. 3/8 in.   d. 1/2 in.

11

Waterproof Techniques
Used With Siding Installation

If anything can be said about the techniques of waterproofing joints between wood siding and other materials, it would probably be that today it is simpler than in years past. This does not mean that techniques used fifty years ago are not used today, on the contrary, they are very much in use. But today vinyl and other types of caulking are available, and these materials are so improved that they retain their elasticity or plasticity for many years. This, of course, means that they may be compressed and stretched as building components expand and contract.

Caulking is effective in sealing joints between wood and wood siding, but these two materials frequently expand and contract at the same rates. This could be severe; however, experience has shown that there is little problem encountered here.

Where caulking is used between brick or block and wood the situation is different. These materials are considerably different and their expansion and contraction rates are quite dissimilar. Wood will expand much more than brick. The unequal changes in these materials place the greatest stress on caulking, so the best nonhardening, nondrying types should be used.

Felt paper either 15 lb. or 30 lb. is an old standby that is used as backing to prevent leaks. It is usually installed horizontally and overlapped in the direction of rain fall. However, felt is frequently omitted from the exterior sheathed wall, especially if bituminous coated sheathing panels are used.

Even over this type of sheathing, felt is used as a waterproof backing in several places, at the corners of a building where strips of it are installed vertically, and behind window and exterior door

243

assemblies where strips are doubled and nailed in place before these units are installed.

The third waterproofing material is also an old standby, metal flashing. It is usually made from three metals: copper, aluminum, and galvanized steel. In special cases zinc is used. Flashing metal is usually purchased in rolls or by the foot from a roll; it is cut on site as needed and shaped to any contours that are required for it to be effective. Two of the most common areas for its use are flashing between side wall and intersecting roof and over the rain cap on wood-framed windows and doors.

WATERPROOFING THE JOINT BETWEEN CORNER BOARDS AND SIDING

Regardless of the type of sheathing used, but especially where insulation sheathing is used, strips of felt vertically installed will seal corners from water that may penetrate the joint of corner board and siding. Figure 11-1 illustrates this technique. A single strip of felt should be cut to reach from top plate to sole plate for each interior and exterior corner of the building. These pieces are then installed as shown in Figure 11-1. Notice that the felt is applied over the sheathing board. This method of using felt develops a fairly wide surface either side of the corner.

Corner boards are often prepared on the bench rather than constructing them on the wall. One piece is usually left at full width (a nominal 1 x 4). The other piece of the corner is usually ripped to 2-3/4 in. wide so that when nailed to the 1 x 4 two finished surfaces equal in width to the 1 x 4 are made.

Where 3/4-in. thick sheathing is used over the wall this size corner board is very effective and of the proper size. It is attractive, not too skimpy or too wide. Further, siding can be nailed easily to studs behind the sheathing. Or, in installation where siding is installed directly over studs (not shown in Figure 11-1), adequate nailing surface is present. Notice the bottom drawing of Figure 11-1a; it shows the amount of direct contact or overlap of the corner board and corner studs. This top view also shows a good nailing surface for siding.

Where 2-in. rigid insulation board is used for sheathing to create a thicker, more thermally insulated wall, felt is still used in the same manner as Figure 11-1b shows. The corner board is installed over this covered corner; but notice in the lower illustration

Figure 11-1 Waterproofing exterior corners where corner boards and lap siding join.

that only a small overlap exists between corner board and stud. Toenailing will be required to hold the corner board in place. It probably would be better to use a 1 x 5 corner board. This extra width would provide a bit more overlap and sounder nailing, and a 1 x 5 corner would not look too wide.

Lap siding is shown butted to the corner board; however, other types, including cedar shingles, may also be used. The type of siding used is not the significant issue in waterproof installation at the joint. It is the way the joint is fitted. For most installations a finely cut close fit is most appropriate. Several master carpenters recommend a slight *under cut* to the siding's end. This ensures a snug fit. As long as each piece is snugly and neatly fitted *and does not* create pressure on the corner board the fit is proper. No caulking should ever be required and normal applications of paint or sealer will complete the seal.

WATERPROOFING THE JOINT BETWEEN SIDING AND BRICK WALL OR BRICK VENEER

It is quite commonplace to use brick veneer on the front wall of a house (see Chapter 2). At other times brick walls intersect with walls constructed with wood. In each of these occurrences a condition exists where two dissimilar materials adjoin. This means that these materials expand and contract at different rates, and that they also react differently to wind and seismic pressures.

Several possibilities exist for waterproofing these joints. To make the joint effective and long lasting, some understanding of the problems and construction techniques of these walls needs to be developed.

Waterproofing Between Siding or Corner Board and Brick Veneer

Figure 11-2 illustrates a typical technique used to make a watertight joint. Carefully examine the illustration to determine the construction of the wall. For example, notice that a 1-in. air space exists between the interior face of the brick and outer surface of the sheathing. This is an acceptable and preferred condition. Next note that the corner board (1 x 4) is nailed as closely as possible to the brick. This arrangement creates a small space where caulking can be applied.

Figure 11-2 Waterproofing the joint between siding and brick veneer.

Caulking is only a part of this waterproofing technique. Felt paper, as illustrated earlier, should be installed under the board and lap around the corner. Any moisture accumulated between wall and brick will fall to the foundation and escape through the weep holes between bricks.

Fitting the corner board to the brick is not usually a difficult task. However, fitting one to a fieldstone or sandstone veneered wall is quite difficult but no less necessary. Scribing is usually required, and once the cutting is complete, caulking is needed to seal the joint.

Waterproofing between Siding that is Butting to a Brick Wall

The brick wall may be part of a room or the side of a chimney or fireplace as Figure 11-3 indicates. The reason for joining the two types of wall is unimportant; what is important is that the joint must be waterproof.

It is frequently necessary to hold the framing away from the brick, but at other times, as when little chance of structural problems or fire exist, the studs are fastened directly to the bricks. Since a space may exist between stud and brick, sheathing is usually installed in line with the studs. Only when the brick wall is installed first will sheathing butt or be cut close to the brick. It is prudent, however, to hold back the sheathing 1/8 to 1/4 in. to allow for expansion.

Thus far no effort to seal the joint has been made. Prior to siding the walls, a double layer of felt may be applied to the intersection. This would be compressed between siding and sheathing and aid in creating a tight joint. A second method would be to apply a generous bead of caulking between sheathing and brick. Following this the siding could be installed and caulking applied as shown in Figure 11-3.

Figure 11-3 Waterproofing a joint between lap siding and brick wall.

The fit of the siding may require the use of a scribe because of irregularities in the face of the bricks. The brick-layers technique of installation may also create imperfect alignment and the scribe may need to be used again for fitting. In either event the fit is sure to be less than perfect, so caulking is a must.

WATERPROOFING THE INTERSECTION OF WALL TO ROOF

One of the oldest techniques of waterproofing uses flashing to create the tightness necessary; but from outward appearances it is not very attractive. Figure 11-4 illustrates the technique used. The roof, in this case a patio, breezeway, or porch roof, intersects with the second-story wall of a two-story house. Although this illustration depicts this situation, remember that all other applications require the exact same treatment as explained here.

Figure 11-4 Applying flashing at a wall and roof intersection.

Before the shingles are applied to the roof and siding is applied over the wall, pieces of flashing need to be cut and folded. It is customary to cut these pieces as long as a shingle is wide (12 in.) and have the piece at least 12 in. wide. Folding through the center creates a piece as shown.

The first piece of flashing is installed overhanging the roof's lower edge by 1 in. *before* the first course or starter course of shingles is laid. The second piece of flashing is laid over the first course of shingles but held back 1-1/2 in. from the lower edge of the first course. This metal is all but concealed when the second row of shingles is installed. Each piece of flashing is laid as previously described until the ridge is reached; there an overlap of flashing needs to be made.

Siding is fitted along the roof line, but *must not* make contact with the shingles. To do so would cause a moisture trap and decay of siding and flashing might result. Rather a 1/2 to 3/4-in. clearance evenly maintained throughout the length of the intersection between siding and shingles is the appropriate way to treat the joint. Nails driven into the siding should be kept as high as possible from the crease in the flashing.

WATERPROOFING AROUND THE CHIMNEY WITH FLASHING

Techniques similar to those illustrated in Figure 11-4 are used when flashing is used around a chimney that protrudes through a roof. Pieces of flashing are cut and installed up the side of the chimney and a single piece of molded flashing is used on the horizontal sections of the roof and chimney line. Figure 11-5 illustrates these pieces.

What makes this installation different from the flashing along the side wall of the house is that two layers of flashing are formed; one is laid onto the roof and bent up alongside the bricks or blocks. The other (counter-flashing) is fitted into the mortar joints between bricks or blocks. Once the flashing and counterflashing are in place, mortar is forced into the joint and this weatherseals the metal to brick. The overlap of the shingles covers the flashing on the sloping sides, but the flashing usually shows along the top and bottom surfaces.

A close look at Figure 11-5 shows that flashing is installed as the roof is being shingled. As soon as the row of shingles that requires a cut out for the chimney is laid, the lower flashing piece is set in place and nailed to the roof. Next the side pieces are installed row by row as the shingles are laid. Finally, the top piece, sometimes called the "saddle cover," is installed. This procedure provides the basis for a watershed off the roof and around and past the chimney. However, counterflashing must be installed next. It laps over the roof flashing and fits into joints that are later tuck-pointed with mortar.

Figure 11-5 Types of waterproofing techniques for flashing around chimneys.

Installing flashing around a chimney is not a simple task. It must be done properly and soundly. Therefore, the skill of a sheet-metal man is frequently used to fabricate the pieces and install them. The mason that laid up the chimney usually is required to tuck point all the joints when the metal is in place.

FLASHING THE HEAD OF A DOOR OR WINDOW FRAME

Metal flashing must be installed over each wooden window and door. Most units are purchased with wooden drip caps installed. If one is not on the unit it is recommended that one be installed. Figure 11-6 shows the flashing installed on a drip cap over either a window or door. Metal, probably aluminum, must be purchased from a local lumberyard to do the job. It is sold by the running foot with widths starting at 12 in. Unless there is a specific reason to do otherwise, a 12-in. wide strip should be split in two because this will give enough metal for two windows or doors. Measure the drip-cap length and *add* 1-3/4 in. when cutting the metal for length. The added length is needed to seal the ends of the drip cap and the wall area to prevent the entry of blowing rain.

The machine edge of the flashing is held even with the bottom of the drip cap. A corrosion-resistant nail (aluminum for aluminum flashing) is driven about midway from the ends through the flashing into the drip cap (see Figure 11-6b). From then on nails are spaced evenly at 1-1/2 to 2-1/2 in. apart; work first on one side of the center, then on the other side until the ends are reached.

Following the nailing, the metal is formed by using a block of 2 x 4 12-in. long and hammer where it is shaped to the slope of the cap and the wall (Figure 11-6b). When the flashing is molded, again working from the center out, press the molding to the cap and wall and nail it near the top into the wall. Four or five nails will hold it.

Figure 11-6c shows the three steps that are used to cut and form the ends of the flashing. The areas indicated for snipping should be cut, then the front piece is folded back. The top piece is then pressed over it and nailed. With all the window and door caps flashed and the building paper installed, the preparation tasks are complete. The layout of siding must be considered next.

Figure 11-6 Flashing the head of a door or window frame.

Drip cap

Window trim

(a)

Nailing sequence

3 2 1   1 2 3

Start here

1½" to 2½"

(b)

Forming flashing to cap

2 x 4 block

Flashing being molded to drip cap and wall

Snip here to wood

Flashing

Drip cap

Step 1

(c)

Trim to make fit

Flashing tucked in

Step 2

Step 3

Use two nails to hold in place

Trim close to make even with wood

Folding flashing over ends

PROBLEMS AND QUESTIONS

1. What characteristics of caulking make it so useful for a building made of different materials?
2. List two common uses for metal flashing.
3. Explain why it is desirable for you to use a strip of felt paper behind corner boards.
4. What type of fit should a lap siding board to corner board be?
5. What would be a reason why studding, sheathing, and siding is kept away from a brick wall or brick chimney?
6. Why are short pieces of flashing most desirable when flashing a shingled roof to side wall?
7. When siding is cut to the slope of the adjacent roof, as in Figure 11-4, rot is minimized if a _____ in. clearance is maintained between shingles and siding.
   a. 1/4 in.   b. 1/2 to 3/4 in.   c. over 1 in.
8. What is counterflashing?
9. Is the saddle cover behind a chimney a single piece of flashing or more than one piece?
10. What condition would have to exist for flashing to be used above a door or window frame?

12

Column Construction for
One- and Two-Story Houses

Columns used to support roof sections and their attached ceilings and for decoration are varied in design and construction. They may be nothing more than a 4 x 4 post to circular columns crowned in doric, corinthian, or ionic orders (Figure 12-1). The modern trend is away from the ionic and corinthian orders; however, the doric order, because of its simplicity, is found in use today. Actually, other titles are given to columns as a whole and these are very limited and somewhat localized. For example, the circular column is generally called a "colonial column," and turned columns milled from 4 x 4, 5 x 5 or larger stock are titled according to style (Figure 12-2).

Thus far the only requirement a carpenter has when using these columns is to trim them for length and install them. There are, however, a variety of columns that are frequently built on site, and these are the ones this chapter develops an understanding of. They are the one-story and two-story built-up types.

ONE-STORY COLUMNS

Cased-Post Column

The *cased-post* type of built up column is one that uses two 2 x 4s as a core. These 2 x 4s are spiked together. This, of course, makes a 3-1/2 x 3 in. post around which the casing is constructed. Figure 12-3 illustrates this type of column, and is used throughout the following discussion.

Figure 12-1 Crowns for columns, Doric, Corinthian, and Ionic.

The reason for using double 2 x 4s instead of one 4 x 4 post for the core is that 4 x 4s are generally sized differently from 2 x 4s. Plus, a 1 x 4 which is used as part of the casing is usually sized to be as wide as a 2 x 4. A close look at Figure 12-3a shows why this is significant. Notice that the casing begins with the fitting and fastening of 1 x 4s to each side of the post. Since both 2 x 4s and 1 x 4s are equal in width, little problem should be encountered fastening these casing pieces. Next 1 x 6s are trimmed in

Figure 12-2 Turned, milled columns. (*Courtesy of E. A. Nord Co.*)

Figure 12-3   A cased column—cutaway view.

width to be equal to the distance from outer surface of a 1 × 4 to the outer surface of the opposite 1 × 4.

Columns of this type may easily be built-up on the bench rather than in place. To do so one must determine several critical factors: (1) What is the overall height of the column? (2) How will the column be attached to the headers or plates of the wall or roof assembly? (3) What adjustments must be included if cornice and soffits are to be fitted around the column? (4) Will the column need a crown? And finally, (5) What base molding is to be used?

The height of the column can easily be found by measuring with a ruler at the place where it will be installed. However, it is much wiser to use two pieces of scrap 1 × 2s as measuring sticks. These are inserted in the place where the column is to be installed then nailed together as Figure 12-4 shows. No height error can occur if this method of measuring is followed.

It may be that the column's core 2 × 4s will extend up into the framing to a plate area, or possibly one of the two 2 × 4s will extend higher into the ceiling or framing. If either situation exists, then additional length must be added to the core 2 × 4s before they are cut. This in no way causes the length of the 1 × 2s to

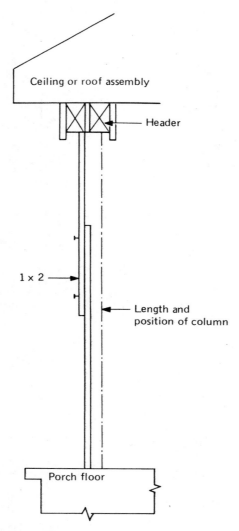

Figure 12-4 Determining column height using measuring sticks.

change, since they are set for the overall height of the column, which should be its casing height. In fact, once the core studs are cut to their required lengths and nailed together, a set of pencil lines should be made around the top of the studs using the 1 x 2 measuring stick as a guide.

The next important step is to determine if all four sides of the column will require casings of the same length. Here several possibilities exist, as Figure 12-5 shows.

Figure 12-5 Fitting details that affect the casing.

In Figure 12-5a the cornice boards, *fascias*, or *friezes* extend an equal distance below the headers. The soffit board is not as a rule extended to sandwich between the column and headers since it is made of materials that are subject to crushing. This situation indicates that the 1 x 4-cased sides of the column will extend the full length of the column. However, the 1 x 6-cased sides will be trimmed shorter than the full length of the column. The amount to be trimmed can be either measured with a ruler or marked on the 1 x 2 measuring stick.

The major difference between the cornices in Figures 12-5a and 12-5b is the width of the outer fascia board. Notice that it is made much wider so that it may be scalloped. This means that each 1 x 6 of casing on the column must be cut to different lengths. This is not difficult to do but individual measurements must be made or a pair of marks must be made on the 1 x 2 measuring stick.

A final factor that may need to be resolved is one concerning the soffit board. If it is cut to allow only the core studs to seat against the headers or plates, then the 1 x 4 casings would need to be trimmed the thickness of the soffit. This is a decision that must be made during the cornice construction and carried through to the column construction.

Suppose, on the other hand, that the column is to be built with a crown; what steps must be taken? Figure 12-6 shows a simply designed crown. First the overall length of the column must be trimmed equal to the thickness of the crown material.

Figure 12-6 Details for adding a crown to a column.

The crown needs to be cut and installed, and cove or crown molding is used in the joint below the crown. The cornice stock must be cut away to allow the crown to make contact with the framing material. One very important characteristic that is easily added to the crown is beveling its top surface on the sides that are exposed to rain. If a slight bevel of 1/8 in. is made on these sides, rain, dew, snow, or water from washing down the house will readily fall away from the surface and joint. This simple alteration eliminates the risk of rotting that so frequently results with crowned columns.

If the column is cased with a 1 x 4 and 1 x 6 as previously described, it lacks a certain formality and has an unfinished appearance. It needs base molding. A simple addition of a 1 x 6 base molding as shown in Figure 12-2 and here in Figure 12-7 provides the final touch.

Porch or entrance floor

Cased post

Flat surface contributes to wood rot

Mitered corners

Figure 12-7 Adding a base molding to a cased column using mitered corners.

Although many different widths of molding could be used, the standard 1 x 6 stock lumber seems most fitting for the single-story column. As shown in Figure 12-7, the molding is cut to fit around the base of the cased post. Obviously a square edge cut could be used; however, four raw edges would show. Painters would need to take special steps to seal them. The results could be good or bad; time would surely tell. Therefore, miter cuts are customarily used. The mitered corner is neat, fits well, and eliminates the raw-edge problem.

The addition of the base molding is not the perfect answer if installed as shown in Figure 12-7, for several reasons. One, it is very square, and second, it contributes to rotting. To be more appealing and to be a deterrent to rot it should have the final appearance of Figure 12-8. The top edge of the base molding should be beveled in the direction of rain (Figure 12-8a). This slight bevel would act as a rain cap and readily shed water from the casing and molding joint. Or a rain-drip molding (Figure 12-8b) could be installed on top of the base molding; it also adds a bit of refinement to an otherwise simple, but nice, base.

Figure 12-8 Beveling the top of the base molding or adding a drip molding to shed water and reduce the chances for decay.

All of the preceding refinements described have been added to a column that is straight. That is, its thickness is the same throughout its length. In most single-story applications this *non-tapered* design is the appropriate one to use. However, suppose a tapered column is needed?

Cased Tapered Column

The two-2 x 4 stud core is still the basis for this column; however, the casing of the post would consist of four tapered pieces of 1-in. stock. In addition, nailing blocks would need to be installed approximately 24 in. o.c. from bottom to top, much like those shown in Figure 12-9a. If a *two-inch* taper were to be built into the column, 1-in. blocks would be installed at the bottom of the column and nailed to the stud core. None would be used at the top. A straightedge laid as Figure 12-9b suggests could quickly define the thickness of blocks at each 24-in. point along the core. Once cut and nailed in place the core and blocks would be ready for casing.

At this point two options are available. One, you can cut the casings for length and mark the centers at each end for width and then measure from the center points to each outward edge a distance equal to *one-half* the casing width as Figure 12-10 suggests. Using a chalkline, snap a line connecting the points previously marked. These lines serve as guides for sawing and planing.

The second option is to use one machined edge of lumber, measure the width of casing from this edge, snap a chalkline, and then cut and smooth the piece as in Figure 12-10b. However, you are now faced with a situation of making a cut at each end of the casing that will properly fit to floor and cornice. This is not easily done for several reasons; namely a cut from the bottom end improperly made causes the entire piece to be either too short or too narrow. A cut from the top causes the piece to be too short and hence too wide as it is raised up.

The method shown in Figure 12-10b can be used, but several things must be done. The length of the casing must be found, along with its width at top and bottom. Then a centerline can and should be snapped with a chalkline. From this line a *right angle* can easily be made with a framing or combination square. Once these are drawn, the single cut for taper can be laid out and all cuts can be made.

A third solution is also possible. Tack nailing the slightly longer than necessary casing onto the modified stud core positions one edge correctly. The opposite edge can be scribed along measurements transferred to its surface from below (Figure 12-11). The

Figure 12-9 Fitting the core with filler blocks for a tapered cased column.

½ width of casing

90°
angle
from the
center line

Chalk line
and cutting
edge

Tapered
casing

Machine
edge

Center
line

½   ½
(a)

90°
angle from
the center
line

(b)

Figure 12-10 Layout of tapered casing pieces—measurement method.

Blocking

Core

Casing

Taper cutting line

A-A transfer points for top end cut

B-B transfer points for lower end cut

C-D transfer points for taper cutting line

Figure 12-11 Layout of tapered casing pieces—overlay method.

top and bottom lengths can also be transferred from the core to the casing. All four casing pieces can be laid out in a similar manner and cut accordingly.

There is one sure way of making each joint professional looking. While dressing (planing) the long-side edge of the casing, slightly undercut each throughout its full length. The nails will draw the boards together beautifully. In fact, if this technique is used, each casing piece should be cut about a full sixteenth of an inch wider than necessary, so, when the bevel is added, the inner or narrower width exactly equals the thickness needed.

Moldings added to the top and base over the casing will not fit accurately if cut square and mitered. They need to be tapered and mitered. This is done quite easily by using a *bevel square*. Once it is set all pieces can be laid out with it making the cross-cutting lines; then a combination square should be used for the 45 degree miter. A word of caution is needed here: Be sure to use the bevel square from the same side of the molding stock each time— all top or all bottom. Switching results in improper fits, open gaps, and frustration, not to mention the waste of lumber.

ANCHORING A COLUMN

It is not sufficient that a column be set in position and left unfastened. The weight of the porch or roof section would, of course, bear upon it to some degree. This is not the solution. It must be anchored properly at top and bottom. The top is frequently held in position by toenailing, although sockets and straps or bolts may be used.

The bottom of the column may be positioned over a previously embedded J bolt, steel pin, or plate and pin (Figure 12-12). In these situations a hole must be center bored into the core so that the bolt may snugly slide into it. This is adequate for most installations, but it does not afford any restriction to keep the column from twisting.

A galvanized metal or aluminum L plate installed under and alongside the core or casing is more rigid and provides sounder anchoring (Figure 12-12b). In this method the wood core, casing, or base molding is chiseled out to a depth equaling the thickness of the flange. Bolts or screws are used to secure the column to the plate. As a rule the plate's vertical flange does not show because the molding or casing and molding are installed over it.

A final touch to complete the bottom of a column is to keep the molding off the floor to allow for air circulation and free passage of water. Most of the time a carpenter will use one segment of his folding ruler as a thickness gauge to keep the molding off the floor. By simply laying it flat on the floor and positioning the cut molded piece on top, he establishes his reference. Another technique that accomplishes the same thing is using a small strip of felt shingle instead of a ruler. After each piece of molding is nailed fast, the ruler or piece of shingle is removed.

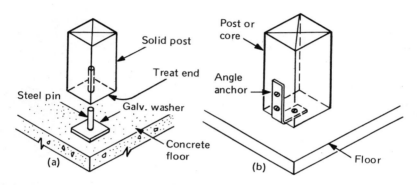

Figure 12-12 Anchoring the bottom of a column using pin or angle bracket.

THE TWO-STORY COLUMN

There is a distinct correlation of principles of construction between the one- and two-story column. The 4 x 4 is not used at all, but the cased-post arrangement is used extensively. The major correlation exists in the techniques used in casing the core, and fastening it to the roof or ceiling assembly and floor. The variations in the core structure may vary widely.

The extremes of core design range from no core at all to an extensive built-up core. The conditions usually prevalent where no core is used is one where the columns are purely decorative and do not support any appreciable load. For example, a small projection of joists built with cantilever construction would be self-supporting. The columns inserted between the lower edge or ceiling of this extension and floor would add beauty and style to the design of the house, but bear little weight.

There is sometimes a set of blocks that are included within the casing as seen in Figure 12-13. These blocks are included to provide nailing surface, but more importantly to maintain the squareness of the column throughout its length. If the column is uniform in width the blocks would all be equal in size; however, if the column is tapered each block would need to be tailored to size at the point of installation.

Figure 12-13 Building a hollow column that uses blocking to maintain column squareness.

If the column needs a core, many different arrangements can be made. However, the overall cross-sectional dimensions would restrict the variety of core assemblies. Two types are shown in Figure 12-14.

The ones in Figures 12-14a and b are made from two 2 x 6s and one 2 x 4 nailed so that an H or I structure is formed. Two different size columns may be made if (1) the 2 x 4 is nailed on edge or (2) flat. If used on edge, an overall 7 x 8 column is possible; with shims an 8-in. square column can be made. If the 2 x 4 is used flat a 6 x 7 column is possible or, with shims, a 7-in. square column.

The third alternative shown in Figure 12-14c uses either two 2 x 6s or a 2 x 4 and a 2 x 6 in an L configuration. In this arrangement blocks similar to the ones shown in Figure 12-13 are used to complete the core. They establish the *square reference* needed to keep the column true throughout its length.

Although more materials and labor are used during the construction of a two-story column with crown and base, there is usually no great difference in principle or task when applying the casing crown and base molding. There is little if any difference in anchoring techniques either. Even the process of constructing a tapered column is the same.

1 Top views of cores
  for cased columns.

2 Add 1½ inches to
  measurements for
  overall column dimensions

Figure 12-14 Alternate core arrangements for larger columns.

What must be different is the cross-sectional dimension. A 4 or 5-in. square column would look very spindly and weak. It would have the appearance of a toothpick holding up a block of wood. Substance and mass are needed. If a group of columns are installed several feet apart, 7-in. columns are adequate. The human eye tends to blend things together and therefore the group of relatively slender columns would take on mass. In contrast, if only four columns are used across a 20-ft. front, each spaced 5-ft. o.c., then columns with larger cross-sectional areas ranging from 8 to 12 inches are more appropriate.

These values should be kept in mind whenever decisions about column dimensions must be made. Along with the decision of column size, the dimensions of its crown and molding base should be selected to add a balanced appearance. Thicker crowns with wider bed molding may add needed mass. Also, the base molding may be more pleasing if 1-1/4 or 2-in. stock are used.

Summary

On-site construction of columns is an exacting task. It re- quires sound applications of the principles of column con- struction, proper fitting, the use of cores in certain cases, and no cores in others. Straightness and squareness are two absolutely essential requirements. All efforts must be centered on these two characteristics because they define the appearance. The carpenter must use all the proper techniques to accomplish these needs, plus add whatever support materials are required.

PROBLEMS AND QUESTIONS

1. Of the three orders of columns, which one is most commonly used today?
2. If a cased column were to be constructed by you and its dimensions were 8 in. square and 9 ft. long what core materials would you use?
   a. 2 × 4s   b. 2 × 6s   Explain why.
3. Why should 1 × 4s be nailed to a core of 2 × 4s before the 1 × 6s are fitted?
4. What are the five critical factors that must be considered when on-site building of columns is done?
5. If you install a crown on a column and do not bevel its top surface, what are you contributing to?
   a. a poor fit   b. a moisture trap   c. excellent craftsmanship
6. Give one good reason for using a miter joint when fitting and installing column base moldings.
7. State the technique you would use on base molding to minimize the chances of rot.
8. Is the difficulty in casing a tapered column the:
   a. Fitting of the casing pieces along the length of the column, or
   b. Cutting the top and bottom of the casing pieces perpendicular with the centerline of the column?
9. What problem can occur with a column if a pin is used as an anchor on the base?
10. What technique is used in hollow-column construction that keeps the column square?

13

Screening A Porch

As this chapter explains, screening a porch is clearly in the advanced skills area of the best carpenter. It requires a fair degree of design capability as well as working knowledge of cabinet joinery.

The method explained in this chapter is not the application of screen wire to 2 x 4 studs spaced 24-in. o.c. It involves the fabrication of panelized frames with joinery and sculptures in such a way that beauty of line and dimension are employed. If done properly, even economy of materials can be incorporated into the design. If done poorly, a considerable waste of materials will occur. If done properly, channels for rain and moisture runoff are included. If done poorly, rot and mildew form immediately.

What then does the overall appearance of this enclosure look like? One example is shown in Figure 13-1. Each panel extends from column to column or wall, and floor to soffit. In this porch each panel is approximately 8 ft. high by 8 ft. long. Actually, each width is slightly different, but as the photo shows there is an apparent unity of dimension. All variations are distributed to the smaller side screen sections. The wider center sections are uniform in width to simplify layout, cutting, fitting, and assembly.

DESIGN

A key element in the planning stage of an undertaking such as screening a porch is the design. Two designs are shown in Figure 13-2. Given a specified panel width, each panel of screen wire can be made equal. This would create a uniform repetitive pattern. On

273

Figure 13-1 Enclosed porch.

most homes this pattern is conservative, adding a dimension of style and harmony. The horizontal member approximately 32 in. from the bottom is included for two reasons: one, it adds a measure of stability and beauty, and it also reduces the chance for damage to screen wire. Should furniture be bumped against the panel or a person walk into the screen by accident only a small panel of wire would be damaged.

The second design shown in the figure is one employing a large 36-in. center panel and smaller side panels. Here the harmony is experienced when adjacent panels are viewed from a short distance. The repetitive pattern is there but not so dynamic as in the first example.

Obviously, many other possibilities exist but care must be used because panels that use wide widths of screen wire tend to look bold and are disproportionate on small houses. However, wide expanses of screen may be appropriate on very large houses. On the other hand, the wooden stiles and rails also add mass to the appearance. Too much wood can be as overpowering as too much screen wire. Therefore, for best results restrict the minimum size of screen width to 18 in. and maximum width to 48 in.

Figure 13-2 A screen panel showing stiles, rails, and overall designs.

Another design feature that must be considered is the thickness of the panel frames. Nominal 1-in. stock is not satisfactory. It tends to be too flexible and skimpy looking. Stock of 5/4-in. thickness is the most appropriate. It measures a full 1 in. to 1-1/8 in. thick. This is stout enough to withstand the joinery used during construction as well as minor abuse, should any be forthcoming.

Still another design feature is the need to build panels 1/2 to 3/4 in. shorter than the actual height of the opening. The space must be included along the bottom edge to allow for rain and other water to escape from the floor of the porch. Lattice strips with inverted Vs cut each 24 in. and at corners are installed over the gap inside and outside to complete the job.

Still another optional feature is the small curve cut into the head rail of each screen section. This gentle arch breaks up the severe straight line. However, not all houses screened with this type of feature in the panel may look good—use caution.

ESTIMATING

Tools Needed for the Job Include

14 oz. claw hammer
framing square
combination square
ruler
#8 hand saw
3/8-in. electric drill
1 set dowel centers
attachments
5/16 in. drill bit
portable power saw[1]

1 set of hand chisels
6-in. jack plane
block plane
4 ft bar clamp[1]
electric (hand) jig saw
   (if curves are planned for)
radial arm saw[1]
miter box with saw
stapler, with 3/8-in. or
   1/2-in. long staples

A Time Table

A time table is developed in the following manner. The number of labor hours based upon woodworking experience should be the judgment of how long it will take to: (1) cut all

---

[1] Optional

pieces for length, (2) rip them if necessary, and dress them with a plane, (3) drill holes for dowels, (4) assemble each panel, (5) install screen wire and put on screen molding, (6) paint panels, (7) install panels, (8) hang screen door, and (9) apply final coat of paint.

Preparing an Estimate

\#_____ 5/4 in. × 8 in. × ____ ft. for horizontal bottom rails
\#_____ 5/4 in. × 4 in. × ____ ft. for chair rails
\#_____ 5/4 in. × 4 in. × ____ ft. for heads of screen panels
\#_____ 5/4 in. × 4 in. × ____ ft. uprights
\#_____ 5/16 in. × 3 ft dowel sticks for doweling joints
\#_____ linear feet of ____ -in. wide screen wire
\#_____ linear feet of ____ -in. wide screen wire
\#_____ linear feet of screen molding
\#_____ linear feet of 3/4-in. quarter-round molding
\#_____ linear feet of 1/4-in. lattice molding

1 box staples	1 pair screen door butt hinges
2 lb. of 6d finish nails	1 lock set
2 lb. of 3d finish nails	1 screen door spring
1 screen door 3 ft. 0 in. × 6 ft 8 in.	
18 linear feet 1 × 2 (stops)	1 lb. of 3/4 in. brads

All quantities and lengths needed for the panels must be filled into the spaces provided.

CONSTRUCTION

The construction phase of the job includes three parts: (1) layout of both porch enclosures and each panel, (2) cutting and assembling, and (3) screening.

Layout

Figure 13-3 shows a typical porch floor plan. This plan is used as an example to aid in familiarity with layout methods. First the position of the screen panels is inserted into the standard porch floor plan. The screen door position is designated against one post

Figure 13-3   Floor plan of porch, defining length of panels.

for sound anchorage of the frame. The length of each panel is measured with a ruler and written into the plan. The results in our
example show one panel 9 ft 0 in. wide, one 10 ft 9 in., one
10 ft 6 in., and one 5 ft 8-1/2 in. wide. The door is a standard 3 ft
0 in. × 6 ft 8 in. screen door.

Now lay out the panels. For this example the 9 ft 0 in. panel
and Figure 13-4 are used. (The others must be made using the
same techniques.)

The panel frame is made of 5/4-in. × 4-in. stock for two full-
length stiles, two intermediate stiles, one top rail and a three-piece
chair rail. The bottom is made from 5/4-in.× 8-in. stock. The finished panel frame must measure 8 ft 11-1/2 in. wide (1/2 in. less
than the opening and 3/4 in. less in height than the total opening
height). These variances are needed to allow for swelling, out of
square, and moisture runoff.

The intermediate stiles are positioned to make three equal
screen openings. The chair rail pieces are marked for position.

Cutting and Assembly

Mass cutting of all the pieces is desirable over individual layout and cutting. All outside stiles and all intermediate stiles can be
cut at one time. Each panel's top and bottom rail should be cut in

Figure 13-4 A panel illustrating doweling, approximate sizes of stiles, rails, and overall height.

pairs. Finally, the chair-rail pieces should be cut and numbered. Actually it is a good idea to number all pieces.

Lay all pieces for one panel on the floor and make sure they fit together properly. Then bore holes for dowels at least 1 in. deep on each part of the joint. Pour or squirt some glue into a dowel hole before inserting the dowel and spread glue to cover the joint.

Start the assembly with the center chair rail and intermediate stiles. Next install bottom rail, then top rail. Follow this by installing one more chair rail piece and then the outside stile. Finally, install the last chair rail and outside stile.

The frame must be square, so a framing square and a 1 x 2 brace nailed diagonally must be used while the glue is drying and while screen wire is being installed. After the glue has dried, sand both sides with a belt sander if you have one, or just plane any irregular joints flush.

Screening

The three upper panels must be screened first, and of these the center panel should be done first. With the wire stapled, the screen molding is installed by cutting and mitering all corners and then nailing with 3/4-in. brads. All excess screen wire is trimmed with a sharp utility knife after molding is in place.

The lower three panels are screened last. If children are at home, a 1/2-in. wire mesh installed over the screen wire and a heavier molding is desirable. *Note:* it is best to prime-paint the panels and molding before installing the screen wire.

PANEL INSTALLATION

Figures 13-5, 13-6, and 13-7 are provided for understanding of the installation of each panel. A 3/4-in. quarter round molding is used alongside each of the stiles and top rail. Lattice, 1/4 in. x 1-1/2 in., is used along both sides of the bottom rail.

1. Cut and nail quarter round molding to post and soffit.
2. Place panel against quarter round and raise to soffit. Toe-nail into posts.
3. Cut and nail quarter round molding to fit against post and stile and soffit and rail.
4. Cut and nail lattice to bottom rail. It must touch the floor. (See Figure 13-7.)

Full side view

Figure 13-5 Full side view of panel in place and molding used to keep it there.

Top view

Figure 13-6 Top view of panels in place showing quarter round on each side of panel.

Side view (base section)

Figure 13-7 Side view—base section. Shows clearance between panel rail and floor and lattice covering space.

SCREEN DOOR INSTALLATION

After the panels are installed, the screen door must be fitted and hung. Figure 13-8 shows a top view of this installation. First, measure the actual width of the door and then add 1/4 in. for clearance. The remaining space between the door and post must be filled in with a piece of 5/4 x 3-1/2 in. stock, ripped narrower if necessary. The 5/4-in. stock should reach from floor to soffit and be toenailed in place very firmly.

Butt hinges make a finer looking job when hanging the door than flush hinges. Above all, the door should open inward wherever there are steps outside. A passage lock needs to be installed to lock the door and keep it closed, and a screen door spring helps pull the door closed.

Normally, the height of the screen door is lower than the soffit, therefore a filler piece of 5/4-in. stock is cut and fitted in this space. An allowance of a 1/4-in. space above the door is essential for proper door operation. Next, the 1 x 2 door stop needs to be cut and nailed in place on the outside, to the panel, the 5/4-in. stock, and the filler piece above the door.

Figure 13-8 Top view of screen door installation. Note filler board between column and screen door.

SPECIAL CONSIDERATIONS

Frequently a panel must fit against brick. If so, caulking can be used to fill the space between panel and brick. If the space is large and quarter round molding is difficult to install a piece of lattice may be fitted in place of the quarter round.

Where a panel must be anchored to a concrete porch floor, angle braces made from aluminum are used. Driving a stud through the metal into the concrete, or setting a bolt into the concrete bolting the angle in place are two methods of anchoring the braces to the floor. Screws through the other flange of the angle hold the panel in place. This should complete the project.

PROBLEMS AND QUESTIONS

1. If screen panels must vary a few inches in width is it better to vary the width of outside panels or the center panels? Explain.
2. List one or more factors that would influence you in designing screen porch panels to choose three screen sections over two sections.
3. Which thickness of lumber is most appropriate for construction of screen panels.
   a. 3/4 in.   **b.** 5/4 in.   **c.** 2 in.
4. What width of stock would yield the smallest waste if your screen panel stiles were to be 2-3/4 in. wide?
   a. 5/4 × 4 in.   **b.** 5/4 × 6 in.   **c.** 5/4 × 12 in.

5. What measurement for panel height versus actual height must be included?
   a. none   b. 1/2 in.   c. 3/4 in.
6. What cabinetmaking technique is used to fasten the pieces of the screen frame together?
7. What type of material is used to close the gap between panel and floor?
8. What clearance is needed on each side of the screen door to allow for swelling?
   a. 1/8 in.   b. 1/4 in.

14

Staircases

A great deal of understanding, knowledge, and skill is needed by any carpenter to properly design and construct a set of stairs. Therefore, this chapter is restricted to defining and examining the principles of stair design for residential use. Residential use means that these stairs will be used by ten or fewer people.

What must be learned are the terms used to describe stairs and their physical location. Providing a written definition of a *stringer* or *carriage* is not sufficient; a clear picture of the stair and its use provides the added dimension that completes the meaning. The principles of stair construction can be divided into two broad categories: (1) those parameters usually used to design residential stairs and (2) the parameters used to prepare proper space for and anchoring of the stairs.

No construction skills and techniques are included in this chapter; they follow in Chapters 15 and 16, where the practical aspects of stair construction are developed.

TERMS

Most of the terms used to describe the parts of stairs are very old, some dating back several centuries. Consider the *carriage* portion of a stair shown in Figure 14-1. This is the principal support member and it indeed acts as a carriage or, more appropriately, as *an undercarriage*. However, the term carriage is not used universally. This same support is also called a *stringer* or *string* for short. Certain segments of the country approve of calling these supports *rough stringers*, *horses*, and even *jacks*.

Figure 14-1 Terms used with stairs and their association.

Whenever a board of finish quality and grade is applied to the outer surface of the carriage or stringer, it may be called one of several names. These are *skirt board, trimmer,* or *stringer board.* Most of the time these are clear-grade stock that are used to cover the ends of the *treads* and *risers* as well as stringers. The *tread* is the board that is normally stepped on as Figure 14-1 suggests. The *riser* is a board installed vertically and under the tread. This means that the outer edge of the tread protrudes about 1 to 1-1/2 in. from the riser. Rather than call this a protrusion, the name *nose* or *nosing* is used. Nosings are irregular curves or form a perfect arc of 180 degrees. On occasion nosing may be added to the tread and may include molding such as cove.

Associated with the tread is the *run of the tread* or *step run.* This is either the horizontal dimension cut into the stringer or is the width of the tread. The *rise* is associated with the riser since the rise is a measure of height from one step to the next.

Most building codes require stairs to be equipped with *handrails*; however, if the handrail is part of an assembly it is frequently called the *balustrade.* Actually, the handrail and the *balusters,* which are the vertical supports of the handrail, form the balustrade. All these are shown in Figure 14-2. Where the handrail ends

Figure 14-2 Terms used with balustrade, baluster, handrail, newel.

Figure 14-3 Housed or closed stringer.

in an open area, as near the bottom step, a *newel* post is used. As a rule this turned post terminates the balustrade.

If a set of stairs is built so that no trimmer is used the stringers are dadoed for the risers and treads; these are called *closed* or *housed stringers* (Figure 14-3).

Another of the terms associated with stairs is one that is used with curved stairs called the *winder*. This is the tread that is cut in *pie* shape. There is also the *landing*, which, of course, is the flat area built whenever the stairs need to change direction.

PARAMETERS FOR RESIDENTIAL STAIRS

Width of Stairs

The usual range of stair width for residential use is from 30 to 48 in. Stairs made narrower than 30 in. are seldom used today. The trend has been for some time to use a 36 to 39-in. stair width. In larger houses, and frequently in split-level houses, wider stairs are used, some as wide as 48 in.

The 36 to 39-in. stair provides sufficient width to make climbing or descending comfortable and for the transport of furniture. This width requires a handrail on one side only. Stairs wider than these require two handrails according to several building codes.

The selection of stair width is usually made by the design architect. He selects it based on available space, purpose, and structural considerations which could, for example, include a landing.

Width and Depth of Landing

Landings are usually required where: (1) more than thirteen steps are needed to reach from one floor to the next. (2) A stair must make a turn. This may be a 90 or 180 degree turn. (3) A door is installed at either end of the stair.

If a landing is needed its width is usually equal to the stairs; a 36-in. wide landing for a 36-in. stair, for example. Its depth (Figure 14-4) is also equal to its width. As the figure shows, a depth of 36 in. is used for a 36-in. wide stair.

If a landing is used where a door opens onto it, its depth must be adequate to allow opening of the door and to allow the

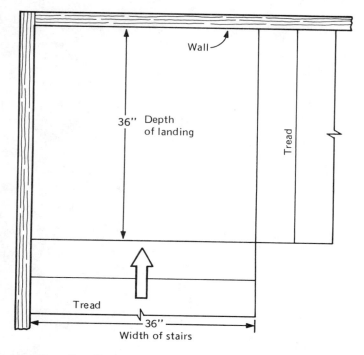

Figure 14-4 Landing dimensions.

person opening the door to stand safely on it. This suggests that a depth slightly in excess of the width of the step is necessary (Figure 14-5). On the other hand, a landing is not required if the door opens away from the direction of the step, provided the rise from the last step to the floor does not exceed the normal step (Figure 14-6).

Where a door opens onto a stair area at the lower level a floor area must be provided; normally this will be framed and floored or treated as a landing (Figure 14-7); however, a landing or floor area should be provided between the last step and door if the door opens away from the step (Figure 14-8).

Rise and Run of Individual Steps

Over a long period of time, carpenters have passed on by word of mouth several dimensions for the rise and run of steps. One of these is the "combination of 17 in." (Figure 14-9). This combination has been translated to many different proportions;

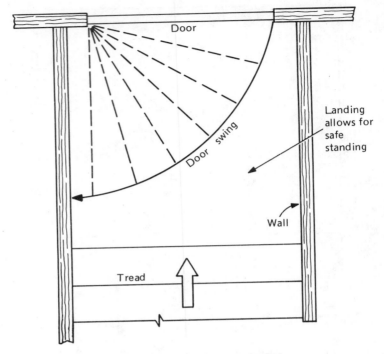

Figure 14-5 Landing where door opens toward the stairs.

Figure 14-6 Stairs where door opens away from stairs—no landing.

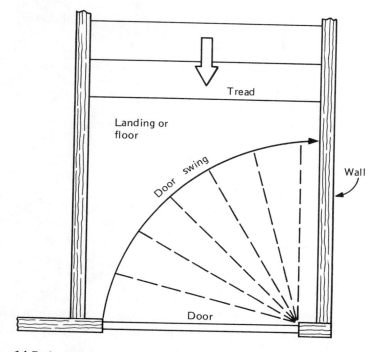

Figure 14-7 Lower landing where door opens toward stairs.

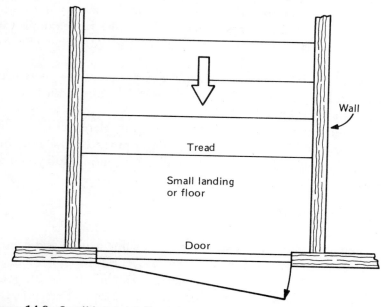

Figure 14-8 Small lower landing where door opens away from stairs.

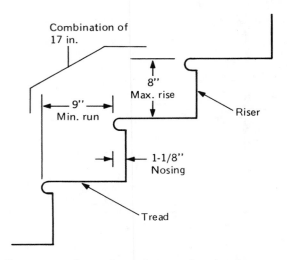

Figure 14-9 Parameters of step rise and run and nosing.

the most common of which is an 8-in. rise and 9-in. run. In actuality the combination of 17 in. meant that a maximum of 17 inches was to be used. Still, this combination allows for a variety of rises and runs to be selected.

Another combination is the "combination of 25 inches." When using this combination, two rises and one run must not exceed a total of 25 inches. This could result, for example, in using an 8-in. rise and 9-in. run for a total of 25 inches (Figure 14-9).

Several building codes have set down the values of rise at 7-1/2 in. and run at 10 in. But, for residential houses, an 8 and 9 combination is acceptable.

The selection of rise and run is not only determined to fall within these combinations but is also determined by the available space for the stairs as well as the total rise from one floor level to the next.

Rise and Run of the Stairs

Where, for example, a standard 8-ft ceiling is used and ceiling joists are 2 x 8 sheathed and floored, a total rise of 96 in. + 7-1/2 in. + 1-1/2 in. = 105 in. is measured. The aim of all good stair design is to have an equal rise for each step. Therefore, the total rise (105 in.) would be divided until an equal number is obtained with the proviso that the 8-in. rise is not exceeded.

In this case if 8 in. is tried first the results show:

- An 8-in. step rise with 105-in. total rise produces the number 13.125 rises.
- Therefore the total number of rises must be increased to 14.
- Dividing 105 in. by 14 rises = 7½ in. per rise.

This method may be used for any total rise measured.

If, for example, a more gentle climb is desired 15 risers could be planned for, with a result that each rise would be exactly 7 in.

However, the decision of whether to use 14 or 15 risers alters the total run. If 14 risers are used there will be 13 runs as Figure 14-10 shows. These runs should be no narrower than 9 in. and may extend in width to 9-1/2 in., in which case the total run is 9 x 13 or 117 in. *or* 9-1/2 x 13 for 123 in.

If a 7-in. rise is used in lieu of a 7-1/2-in. rise, 14 step runs are required. Again, not exceeding the 17 combination, a step run could range from 9 in. to 10 in. This translates to a total run of 9 x 14 or 126 in. *or* 10 x 14 or 140 in.

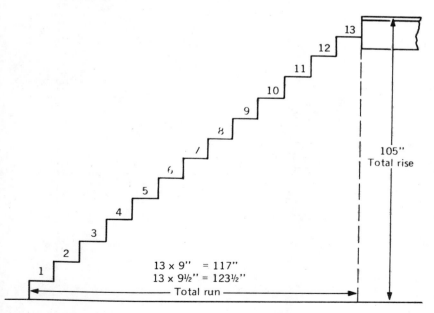

Figure 14-10 Determining total rise and total run of a stairs.

The various combinations shown here allow considerable flexibility of design by the carpenter. If there was adequate room available, a stair with a 6-5/8-in. rise and 10-3/8-in. run could also be constructed. However, there is another factor that may be considered in stair design and it is the angle of the incline.

Angle of Stairs

All stairs used in residential houses should have an angle of slope between 30 degrees and 40 degrees. Figure 14-11 shows both extremes. Here are three cases:

1. The 30-degree angle for a typical total rise of 105 in. results in a wide step of about 11-1/2 in. with a rise of 7 in.
2. The 35-degree angle for a typical total rise of 105 in. results in a step about 10 in. wide with a rise of 7-1/2 in.
3. The 40-degree angle for a typical total rise of 105 in. results in a rise of 8 in. and a step of about 9-1/2 in.

Figure 14-11 The steeper the angle the steeper the stair and less run distance required.

Number of Risers

There is no hard and fast rule for the minimum number of risers per staircase, but general guidelines and some building codes limit the number from 13 to 15. This means that if a situation arises where more than 15 treads are used, a landing should be built somewhere along the travel of the stairs to break up the long climb. Fortunately, most modern residential construction uses an 8-ft ceiling, so use of a landing for the purpose stated above is not needed. If, however, a 10-ft ceiling were used, a check with local building codes would be required to see if a limitation were included. Then, too, the architect may include a landing in the plans for the stairs and thus eliminate the problem.

Handrails

In principle, a handrail should be included whenever a stair is built. Where the stair is installed between walls a single rail installed 30 to 34 in. above the nose of the treads (Figure 14-12) is required and should extend slightly below the bottom step and past the top step. If either side of a stair is open it should have a handrail as part of a balustrade assembly. Stairs that have both

Figure 14-12 Handrails should be 30 to 34 in. up from the nose of the step.

sides open must have a balustrade assembly on each side of the stairs.[1]

Many split-level homes have situations where short segments of stairs using four or fewer risers are made. If this is the case, no handrails need to be installed for safety; they may be installed for esthetic value. But where these stairs use more than four risers, handrails are usually required.[2]

PARAMETERS FOR PROPER HEAD CLEARANCE

There must be a proper amount of head room for people using the stairs, and there must be a *positive feeling* by all persons using the stairs that there is adequate head room. The first is clearly defined in plans and design of the stairwell and if constructed properly adequate head room is provided. The second, the feeling that there is adequate head room, can also be satisfied if proper design and construction are followed.

Figure 14-13 Framing the stairwell.

---

[1] "Stairways," Uniform Building Code (1973) SEC 3305, p. 480.
[2] Ibid., p. 481.

Defining The Stairwell Size

Most residential stairs require a stairwell, or opening, approximately 10 ft long by the width of the stairs (Figure 14-13). Several factors of vital importance can confirm or refine the accuracy of the 10-ft general opening size.

First, the vertical head room that satisfies both the *requirement and feeling* is 6 ft 8 in. taken from the nosing of the second step, as shown in Figure 14-13. This head clearance must be maintained for the remainder of the rise of the stairs. The stairwell length, therefore, should be equal to the total run of the stairs minus the width of the first tread. This dimension is a variable before a design is selected since the rise and the run of the step determine the total run as was explained earlier. However, once a design is selected a fixed length can easily be found. The following problems illustrate the point:

Problem 1: The stairwell opening for a stair rising a total of 105 in. using a 7-1/2-in. rise and 13 step runs is illustrated in Figure 14-14. Determine the length of the stairwell.
Solution:
1. Measure down from the ceiling a distance 6 ft 8 in. and draw a horizontal line.
2. Since the line crosses the surface of the second tread, draw a vertical line up from its nosing thru the ceiling joist.
3. The finished distance for the stairwell is 9 ft 6 in. The rough opening must be made larger to account for trim, riser, and a portion of stringer where the nosing or flooring could be nailed.

Problem 2: In this problem the stairwell opening should be longer than the one in problem 1 since the total rise is the same (105 in.), but the rise per step is reduced to 7 in. and the treads (run) are 11 in. (Figure 14-15). Determine the length of stairwell.
Solution:
1. Measure down from the ceiling a distance of 6 ft 8 in. and draw a horizontal line.
2. Since the line crosses between the second and third tread the finished stairwell opening should be taken as perpendicular from the nose of the second step, even though a slightly greater distance for head room is made.
3. The rough opening for the stairwell should be made approximately 12 ft 6 in. to allow for trim of the stairwell, and a 6-in. portion of the carriage or stringer.

Figure 14-14 Defining the *finished* and *rough* opening length of a stairwell (problem 1).

Conclusions

The general statement that a stairwell opening should be 10 ft long holds true for stairs whose rises are 7-1/2 to 8 in. and have treads of 9 to 9-1/2 in. However, greater head room clearance is needed when a stair is designed with risers less than 7-1/2 in. and/or treads of 11 in. In this case the guidelines described in problem 1 or 2 should be followed *before* framing the stairwell.

If the ceiling over the stairs slopes at the same angle as the stairs as in Figure 14-16, head room clearance must be maintained from the second step to the top step. As long as 6 ft 8 in. is

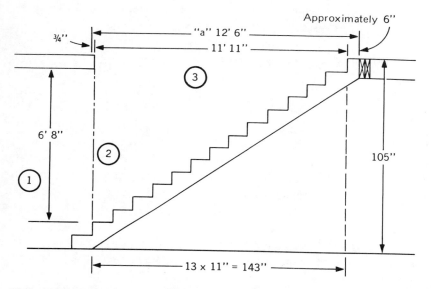

Figure 14-15 Defining the *finished* and *rough* opening length of a stairwell (problem 2).

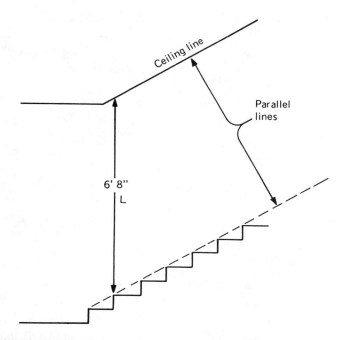

Figure 14-16 Minimum head clearance for main floor to second floor stairs is 6 ft 8 in.

established and is then maintained, both proper design and a positive feeling by users of the stairs is accomplished.

Variance for Head Clearance for Basement Stairs

The previous examples were developed for stairs in the living quarters, as between first and second floors. The basement stairs are generally constructed with the same principle of rise and run, but the allowance for head clearance is reduced to 6 ft 4 in. There does not seem to be any apparent reasoning for this reduction in allowance except tradition. There may be local building codes, however, that restrict this reduction in head clearance, so a check with the local building inspector should be made before construction starts.

As a general rule, basement stairs employ a full 8-in. rise and 9-in. run per step. The parameter, of course, reduces the length of the stairwell. However, even when using a 6 ft 4 in. head clearance the edge of the stairwell falls (using a perpendicular line) between the noses of the second and third step.

ANCHORING TECHNIQUES FOR CARRIAGES OR STRINGERS

A variety of techniques is available to choose from when anchoring a stair carriage or stringer is needed. It could be:

1. Toenailing the upper end of the stringers to the headers
2. Installation of a ledger nailed to the header and notched stringer
3. The use of steel angle braces lag bolted to both header and stringer
4. Anchoring the closed carriage to the side wall by nailing it to studs
5. Nailing a rough stringer to a previously installed trimmer board
6. Inclusion of a kicker plate in back of the first riser board
7. Any combination of these techniques (some are illustrated in Figure 14-17).

The selection of anchoring technique or combination of techniques must be made on site. It should be one that provides several

Kicker plate
(anchor)

Kicker plate
technique

Rough
stringer
nailed thru
finish stringer
into studs

Finish
stringer

Stair carriage

Riser

Nail

Tread

Anchoring to
wall studs
technique

Floor

Base

Header

Ledger

Stringer

Ledger anchoring technique

Figure 14-17 Anchoring techniques for stairs.

important features to the stairs. First, the technique used must provide safety, so that no possibility of stair collapse can ever occur. Next it must provide rigidity so that absolutely no feeling of sway, give, or motion is ever felt by persons using the stairs. In addition, the technique must be planned into the construction and used in that manner during the stair construction. The anchoring device used should not show in the finished product. However, if it must show then it must appear as a fundamental part of the stairs and not look like an add-on. In every set of stairs built with open stringers or jacks, the kicker plate will probably show unless, of course, a trimmer is cut and laminated over the rough stringer.

Many of the techniques listed before are easily used on the closed carriage type of stairs since ledgers, kickers, and even rough stringers are completely hidden. The finished product incorporates safety and displays expert craftsmanship.

Summary

The principles of stair design and construction are varied and many. Those described and explained in this chapter are the most often used in the construction of straight stairs. Even a set of stairs with a landing is really two sections of straight stairs. Risers ranging from 6-1/2 to 8 in. must be used along with a step run of 9 in. or more if good design is to be used. Further, quality materials must be used for rough stringers, carriages, trimmers boards, risers, and treads. Then too, each step tread should have a nosing that extends beyond the face of the riser board by 1-1/2 in. Finally, sound anchoring techniques must be designed and built into the stairs for safety and lasting quality.

PROBLEMS AND QUESTIONS

1. List three other names for a stair's carriage.
2. What is a trimmer board? What other names may it be called?
3. What is the difference between a "riser" and "rise"?
4. What is nosing? How much does it usually overhang the riser?
5. What parts make up a balustrade?
6. What is the range of stair widths for a residential building?
7. If a set of stairs is 39 in. wide how long should a landing be?
   a. 39 in.   b. 45 in.   c. 60 in.

8. Explain what a "combination of 17 in." and "combination of 25 in." mean in stair layout.

9. What is the maximum rise allowable for residential stairs? Minimum stair tread?

10. If the total rise from floor to floor is 108-1/2 in., what would be the average rise if 8 in. is the maximum allowable rise per step?
    a. 7-1/2 in.   b. 7-3/4 in.   c. 8 in.

11. Which angle of stair incline produces more stress on the person using it?
    a. 30 degrees   b. 40 degrees

12. A handrail must be installed between 30 and 34 in. above the nosing of the tread. True or False?

13. Two different allowances for stair head clearance are established:
    a. For main floor stairs it is _____ .
    b. For cellar stairs it is _____ .

14. If a set of stairs has thirteen step runs and each one is 10 in., the rise is 7 in. per step, and the head clearance must be 6 ft 8 in., what is the length of the stairwell opening?

15. List four different anchoring techniques for stairs.

15

Interior Stairs

This chapter details the planning and layout of basement and formal stairs. These two stairs may differ not only in appearance but in construction techniques.

BASEMENT STAIRS

Layout of Stringers

Basement stairs should be constructed with 2 x 12-in. stringers or carriages and 2 x 10-in. treads. Frequently, there are no riser boards used on these stairs. The example chosen for this section is a set of stairs using roughly an 8-in. rise and 9-in. run. The total height (rise) is equal to:

ceiling clearance	7 ft 11 in. =	95 in.
joist width		7-1/2 in.
subfloor and flooring		1-1/2 in.
Total		104 in.

First Trial

Fourteen risers are most commonly used for interior stairs, but in basements where headroom is sometimes lower, as in this example, 13 risers at 8 in. exactly equals the total height.

13 risers x 8 in. per rise = 104 in.

Next is the layout of the carriages. For this a framing square is used.

1. The 2 x 12 should be *sighted* and marked for its crown.
2. Since a 2 x 10 actually measures 9-1/2 in. the step run should be set at 9-1/2 in. (stairwell opening permitting this latitude of variance).
3. Allowance of nosing (overhang of tread) is set at 1-1/2 in. (Figure 15-1).
4. This means that the 8-in. mark on the square's blade is used for the step run. (8-in. seat for the tread and 1-1/2-in. nosing (overhang) = 9-1/2-in. tread width).
5. The 8-in. mark is used on the square's tongue for each rise (6-1/2-in. carriage use + 1-1/2-in. tread thickness = 8-in. rise).
6. Thirteen risers are laid out in order, starting away from the end of the 2 x 12.
7. The bottom cut is laid out next (Figure 15-1). Notice that the thickness of the tread (1-1/2 in.) is subtracted from the rise so that when it is installed the first step is 8 in. high. The bottom cut must be parallel to the first tread run.
8. The top step run is measured for depth to allow for 1-1/2-in. nosing, tread overlap on header if any, and rim of carriage. The sum total cannot exceed 9-1/2 in.

Therefore, assume a 1-in. lap onto the header for the tread. This causes a reduction of the top step carriage run of 1 in. or a total of 7 in. (Figure 15-1b).

The stringer should be carefully cut along all the lines drawn and then another one can be laid out by using the first as a pattern. It too should be cut out.

Installation of Stringers

The first stringer to be set in place is the one against the wall, assuming one or both sides of the stairs is against a wall. It should be placed with its top against the header and positioned properly. An 8d nail should hold it in place temporarily.

Before any final fastening is done, a spirit level must be laid across the step run to verify the levelness of the run. If the level

Figure 15-1 Basement stair stringer layout using an 8-in. riser and 9-in. tread with 1-in. nosing overhang.

indicates an out-of-level condition the bottom must be marked and trimmed if too high or shimmed if too low. (Figure 15-2). There really is no surprise to experienced carpenters that alterations are made. The concrete floor of most basements is not uniformly flat, plus the stringer was cut for an ideal condition.

The next operation is to determine the anchoring technique and apply it. Recall that a kicker board could be used at the bottom and nails could be used through the stringer into wall

Figure 15-2 Installing stringer against wall after level adjustments are made

studs. Toenailing or angle brackets and lag screws could attach the top of the stringer to the headers.

The second stringer is set in place next. First it is set to the proper height and spaced properly for width at the headers. Then it is tack nailed with an 8d nail. Next, a straightedge is placed across the two stringers near the lower end as Figure 15-3 shows. The spirit level is laid on top of the straightedge and adjustments to the second stringer are made.

Once the adjustments have been made, spreader sticks cut to the exact width of the stairs should be nailed across the face of a lower rise and middle rise. These will help maintain alignment while the treads are being installed.

Figure 15-3 Installing the outer stringer, leveling and spacing it, and install-ing treads. Note spreader to aid in maintaining width.

Installing Treads and Risers

The sample set of stairs laid out in Figure 15-1 is designed for treads only, so the first operation after installing the stringers is to install the treads. Assuming that they are precut to length from 2 × 10 stock all noses should be rounded slightly with a block plane or router. Next, starting from the lowest step, each tread is posi-tioned and nailed with face nails (casing type) through the tread into the stringer. Since no risers are used each tread should butt to the rise cut on the stringer. The final or top tread should lap the header by 1 in. and be nailed to the header as well as stringer.

The temporary spreaders can be removed once the treads are in place. For practical purposes the stairs are complete. However, if one side is open it must have a balustrade. If neither side is open a handrail mounted on brackets fastened to wall studs is used. For the description of installing a balustrade skip to the end of the next section on formal stairs.

FORMAL STAIRS (FULL FLIGHT)

There are two distinctly different methods of constructing formal stairs for use between the main or first floor and second floor of a residential house. The first requires the use of the *dado* or more precisely the *blind dado* joint. Considerable skill is needed for layout as well as cut out and assembly. However, stairs constructed in this manner use the closed carriage and a minimum number of pieces.

The second method uses a variety of pieces and has as its foundation a pair of rough jacks. These are completely encased by trimmer boards, risers, and treads. Frequently, this method results in more work than the one previously identified because of all the precise cutting and fitting that must be done.

Laying Out and Dadoing the Closed Carriages

Rather than reiterate the formal description of determining total rise and run and performing all the calculations of how many step rises and runs, assume that a 7-1/2-in. rise and 10-in. run plus 1-1/2-in. nose are planned. Assume to that 14 risers are needed.

The first requirement is to obtain clear stock 2 x 12s of fir or pine, preferably southern pine, for use as carriages. The grade must be the finest available for appearance and strength since no rough jacks are to be used. Layout begins with striking a line 1 in. to 1-1/2 in. in from the board's edge as Figure 15-4 shows. Then the framing square is used to lay off the rises and runs of all the steps. Next, the shape of the nosing is drawn so that its extreme edge is equal to the width of the tread (11-1/2 in.). The one shown in Figure 15-4b (step 3) is half found. Then connecting lines are drawn as shown in step 4. The layout in step 4 also includes a

Rise
7½"

Run less
nosing
10"

1" to 1½"

Step 1   Step 2

Step 3   Step 4   Step 5

11½"

Nosing
layout

Tread and
wedge layout

Riser and
wedge layout

11½"

10"

Alignment
line

5/8"

Tread pattern
with wedge

Alignment
line

7½"

5/8"

Riser pattern
with wedge

Figure 15-4 Layout sequence for closed or housed stringer templates aid and simplify layout.

layout for a wedge. The last step is shown in step 5 where the riser and wedge are laid out. If these series of steps are followed an accurate layout is easily accomplished.

However, some simplification of the layout task can be made if templates for tread and wedge, and riser and wedge are made (Figure 15-4). These make the work much easier. If the templates are used, steps 1 and 2 are performed first. Then the alignment mark on the tread template is held evenly with the rise mark and the pattern is drawn. The alignment mark on the riser template is held evenly with the tread (run) mark.

Once the layout is complete the upper and lower markings are made and the housed stringer (Figure 15-5) is ready for cutting and dadoing. When completely cut and set in place it should resemble the segment shown in Figure 15-5b. Wedges as shown should be cut and made ready.

Normally, the housed carriages are *not* installed before the stairs are assembled because of the difficulty involved. Rather they are laid on the floor and assembled there. Figure 15-5c shows how the assembly is made from the underside of the stairs. Glue is applied to the dado, then a tread is set in place and a wedge is driven behind the tread. This action drives the tread firmly against the dado's edge. Next the riser board is set in place and wedged and the lower edge of the riser is nailed to the tread. Once all pieces are installed into both carriages, glue blocks are installed as shown.

This type of staircase is usually used where both sides of a stair are enclosed with walls. Naturally, some small allowance, approximately 1/2 to 3/4 in., is needed to make installation possible. This clearance can easily be covered with a small cove molding once the stairs are in place. Handrails are usually fastened to one or the other walls 30 to 34 in. above the stairs.

Stairs with Fitted Trimmer Boards and Mitered Risers

The other type of formal interior staircase is made with rough stringers, treads and fitted risers, and trimmer boards. Figure 15-6 shows such a staircase. A primary need to construct this type of stairs is found where one side of the stairs is against a wall and the other side is open and uses a balustrade.

(a) Penciled layout before dadoing

Rise 7½"

1½"

10" Tread (run)

Nosing

Housed stringer

(b) Dadoed housed stringer

Wedge for tread

Wedge for riser

Profile of riser, tread and wedges

(c) Segment of completed stairs with wedges and glue blocks

Blocks glued and nailed

Glued wedges under treads and behind risers

Housing rabbeted to depth of 3/8" to 1/2"

Figure 15-5 Layout, dadoed stringer, and wedges and a section of the glued and nailed stairs.

312

Figure 15-6 Using an inside trimmer board and rough stringer against a wall.

The construction of this type of staircase is straightforward. Rough stringers are laid out and cut. Next a trimmer board is cut and installed and then one of the stringers is nailed in place against it. Before nailing the second stringer in place it should be used to lay out the trimmer board that will be nailed to it. The line drawn for the tread run on the trimmer board is the cutting line; but, the line drawn for the riser is the interior mark of a 45-degree miter. This means that a 45-degree layout must be made on each rise and then cut (Figure 15-7). The risers are cut with one end square and the other at 45 degrees to match the trimmer board. This arrangement makes a professional job with no raw end wood showing when assembly is complete.

When the trimmer board is ready, assembly should begin by installing the rough stringer and trimmer board. The first riser needs to be cut to fit and trimmed for width as well. Glue should be used on the miter joint and on the faces of the stringers for lasting hold and quality. Next the tread needs to be prepared and

Figure 15-7 Using an outside trimmer cut to fit against a rough stringer. Mitered risers fit to 45-degree bevels made in trimmer.

installed. Since one end of the tread is to show, its raw edge must not be seen and a self-return is used (Figure 15-8). Actually, all treads should be prepared at one time. Notice also that the return stock extends about 3/4 in. beyond the riser. The return stock should be glued as well as nailed.

Installing the Balustrade

Following the installation of all risers and treads the balustrade is installed. This, recall, requires use of a handrail and balusters. Normally, two balusters are used on each tread where they are toenailed and glued once their centering position is determined. The upper end of each baluster is tapered to fit under the handrail and there should be a short and long baluster per tread.

What is a simple way of defining the bevel cut needed for the top of each baluster? Line up a spirit level under or alongside the trimmer board and set a miter square against the level with the blade along the trimmer. Then tighten the wing nut. Or, set a framing square on the floor and let its tongue rest against the stair's trimmer board. Place the miter square against the square's tongue and tighten the wing nut when the miter square's blade is properly resting against lower trimmer edge.

Figure 15-8 Self-nosing stair tread.

Applying the handrail to the balusters is the most difficult part of the task; how can it be done easily? One method makes use of a board 3/4 × 1-1/2 in. where the balusters are glued and nailed to it. Top nailing is used. Then the handrail is glued and screwed in place on top of the 1-1/2 in. board. Screws are installed from beneath the 1 × 1-1/2 in. stock.

Another method uses dowels, where a dowel is inserted into the top of each baluster and its other end is inserted into a hole drilled into the undersurface of the handrail. Since the lower end of the handrail and stairs opens onto the floor area a *newel* post is used to finish off the stairs. This post needs to be anchored very firmly. Its rigidity is absolutely essential. Therefore, toenailing is not appropriate by itself. The post should set firmly on the floor (Figure 15-9). A notch into the tread is surely called for or possibly a mortise needs to be made through the tread. The newel post is cut to fit into this opening and anchored to the stringer with lag screws, glue, and nails.

Joining Baseboard to Trimmer Board

The baseboard ends against the trimmer board as a rule. Baseboards in the past were fairly wide and the intersection of trimmer to baseboard was a special joint. Figure 15-10 shows this situation, and it should be noted that a miter is used so that no end wood of either the trimmer or baseboard shows. This technique is still used

Figure 15-9 Several methods of installing newel posts that are sound.

Figure 15-10 Properly joining baseboard to trimmer board at the bottom of stairs and top of stairs—baseboard is the squared 1 x 5 or 1 x 6 variety.

where baseboard is used on landings. The modern trend of baseboard design is the *dustless* shape, where the top edge is curved. Its size is relatively small and its thickness is reduced to less than 3/4 in. This means that it is impossible to join the trimmer and baseboard in the same way. Figure 15-11 shows the modern trend. Unfortunately, some raw edge of the trimmer shows, making additional work for the painters. At the lower end a straight cut is used, but at the upper end the trimmer is trimmed to make the joint more pleasing.

Figure 15-11 Joinging baseboard to trimmer and trimming trimmer board when new *dustless* baseboard is used.

Summary

The building of a set of stairs whether informal or formal requires a great deal of fundamental understanding as well as skill. The formal stairs must integrate with walls and trim and be constructed accordingly. This presents many questions and problems for the carpenter. He must select the correct answer to a problem from a variety of techniques. If both sides of a staircase are walls, then a housed carriage design is appropriate, yet this design requires cabinetmaking skill. If one or both sides of a staircase are open, mitering of riser and trimmer is needed and a balustrade is employed. These decisions result in discrete skills and special techniques.

PROBLEMS AND QUESTIONS

1. What is the maximum height for a set of stairs if the rise is exactly 8 in. and there are 13 risers?
2. Why should you sight a stringer 2 × 12 before beginning the stair layout?
3. What inch marks would you use to lay out the stringer if the rise is 7 in., the tread is 9-1/2 in., there is no riser board, and the tread overhangs the riser cut by 1 in.?
   a. 7 and 8-1/2   b. 7 and 9-1/2
4. Why is the bottom rise on the stringer trimmed for the thickness of the tread?
5. Briefly explain what to do if, when installing a stringer, the tread surface tilts up at the front edge.
6. What is the purpose of a *spreader* when installing a stringer?
7. Using Figure 15-4, explain what physical task is being performed in each step.
8. What is the purpose of putting a wedge under the tread and one behind the riser?
9. If one side of a formal staircase is open (not against the wall) what must be done to the riser cut of the trimmer?
10. What is a self-return on a tread used for?
11. Briefly explain how you would obtain the bevel angle that is necessary at the top of each baluster.
12. Explain one of the three ways to soundly anchor a newel post.
13. When should a *butt* joint be used when installing baseboard against trimmer or carriage? When should a miter joint be used?

16

Staircase for Deck to Ground

Constructing a set of exterior stairs employs the principles identified in Chapter 14 as well as some modifications to the stringers because of extended step runs and more gentle rises. In addition, wood decay becomes a critical factor. Special attention must be paid to this problem to minimize decay and extend the life of the stairs.

PREVENTING DECAY

Wood decay is caused by minute plants called *fungi*. Some merely discolor the wood of a stairs, but decay fungi destroys the fiber. Fungi can only grow in wood when it contains more than 20 percent moisture. Air-dried wood regularly contains 19 percent moisture, which is below the danger point. As long as air freely circulates around the wood the moisture content can remain below the danger point. However, other conditions such as dew, moisture from humidity, and rain all tend to raise the moisture content and contribute to the growth of fungi and to decay.

The conditions that cause fungi to grow and multiply can be eliminated or controlled by more than air circulation. The heartwood of many species of trees is resistant to decay as well as wood borers such as termites. Douglas fir and southern pine heartwood are classed as moderately resistant. Redwood, most cedars, and a few species of red cypress are classed as highly resistant, and these can be in direct contact with soil. However, with time even heartwood decays and obtaining heartwoods is sometimes difficult.

Treating Lumber

The current trend in retarding decay is the use of treated lumber. Treating lumber by painting its surface does not as a rule prevent decay, since moisture may have been trapped under the paint. The application of one of several chemicals is more effective. Some of these chemicals, listed in Table 16-1, may be applied with brush or sprayer, while others are applied under high pressure, or used in vats where beams are dipped.

TABLE 16-1   Preservatives and Application Methods

Type of Preservative	Method of Application
Water-borne salt solutions	pressure
Creosote and creosote coal tar solution	brush, pressure
Pentachlorophenol	pressure
Copper naphthenate	brush, dipping

The salt preservatives leave the wood comparatively clean, paintable, and free from odor. Creosote, on the other hand, blackens the wood; it cannot be painted over and it retains a strong odor for many years. Copper naphthenate turns the wood green and when air dried the treated wood may be painted.

Drainage

Stairs exposed to the elements can last a long time if proper drainage around them is planned rather than left to chance. First, the slope of soil around the stairs should be uniform at sufficient angle to afford complete drainage. This eliminates the chances of pools of water and the development of fungi that might occur. Light or thin shrubbery rather than dense kinds should be planted next to stairs so that air readily passes through and humidity is not trapped near stringers and treads. The bottom of the stringers should not be in contact with soil if possible. But if soil contact is unavoidable, pressure-treated stringers should be used.

One sure way to avoid contact with soil is to have each stringer rest on a block of concrete or pier made of stone, brick, or concrete as in Figure 16-1. When constructing the footing for the

Figure 16-1 Preparing a pier for proper drainage and stringer anchoring.

stringer to sit upon, every effort should be made to slope the edges of the pier so that water sheds away from the stringer and does not become trapped between it and the footing. The one shown in Figure 16-1 has a *mortar* cap, which is shaped to provide the slopes needed for good drainage.

Summary

Proper air circulation is fundamental to the prevention of fungi. Treated wood used for stringers and treads is another and is in common use since heartwood is hard to obtain. Even so, the lower ends of stringers should not be in contact with the soil. Piers are recommended to be used for stringers to rest upon.

STAIRS WITH ROUGH STRINGERS AND 2 × 6 TREADS

The exterior staircase used as an example in this section is one that extends from the ground to a deck. The deck is made of a wood frame with 2 × 4 planking spaced 1/4 in. to allow proper drainage and air circulation. The same principles must be applied to the stairs. In addition, the stringers and treads will be closer to ground level, so treated lumber must be used to resist decay. The use of a cement block as shown in Figure 16-1 is planned to be used as a pier for each stringer.

Objective: Lay out and construct a set of stairs from ground to join a deck 3 ft above the ground. Each tread area is to be made from two 2 x 6s with a 1/2-in. space between them.

Procedure: (Figure 16-2)

1. The step tread width is defined in the objective as 2 times the width of a 2 x 6 plus 1/2-in. air space which equals:

$$
\begin{array}{lll}
2 \times 5\text{-}1/2 \text{ in.} & = 11 & \text{in.} \\
\text{air space} & = & \underline{1/2 \text{ in.}} \\
& \text{Total} & 11\ 1/2 \text{ in.}
\end{array}
$$

2. Assume that the outer tread extends beyond the face of the rise by 1 in. This means the tread run on the stringer is 10-1/2 in. (Figure 16-2a).

(a)

Top

(b) Stringer layout

Allowance for pier

Allowance for treads

Bottom cut layout

Figure 16-2 Stringer layout with bottom allowances shown.

3. Layout of the risers is next (Figure 16-2b).
   a. With a total rise of 3 ft and the greatest rise per step set for 8 in. an unequal distribution occurs. (*Not Acceptable*).

   b. Using 5 risers for a total rise of 36 in. results in a rise of 7-1/4 in. per step. (*Acceptable*).

36 in. ÷ 5 = 7-1/4 in. L

   c. The layout is made as shown, then modified to account for the block and its cap that extend above ground level (Figure 16-2).

4. Should the top rise be a part of the deck framing or part of the stringer?

   a. If the deck has a 2 × 8 header and joist framing with 2 × 4 decking, the total width equals 9 in. The option of using the framing as the fifth riser as Figure 16-3a shows means that *none* of the stringer makes contact. It is necessary to install a separate 2 × 8 header as shown so that the stringer can be fastened to it. If this option is selected the final cut of stringer is made where the fourth riser ends.

   b. If a 2 × 10 frame with 2 × 4 decking is used only 2 in. of bearing is obtained when the deck framing is used as the fifth riser. In this situation, a 2 × 6 header as shown in Figure 16-3b is added to provide more surface contact. As in the previous solution, the stringer is cut at the end of the fourth riser.

   c. The third solution is shown in Figure 16-3c where the fifth riser is a part of the stringer. Here no added framing is added to the deck. The stringer's top edge is held even with the deck's framing and the treads are flush with the 2 × 4 decking.

5. Install the stairs (Figure 16-4).

   a. With all stringers cut and treads precut for finished length the assembly can begin.

   b. One pier block with mortar applied as shown in Figure 16-1 should be prepared for placement under each stringer. A bolt or bent pin can be set into the mortar before it dries.

   c. A small hole should be dug that is large enough to accommodate the block, and it should be set in place. Then the stringer is placed with its top edge properly in alignment against the deck's header.

   d. Using a level, the tread (run) surface of the stringer is made level and adjustment to the block is made accordingly. At this time the drift-pin hole in the stringer bottom should be marked and drilled.

Figure 16-3 Headers added for supporting stringers (a and b) and deck framing as a support (c).

 e. When ready the stringer should be permanently fastened to the header. At this time there should also be a check made to verify that the stringer is at 90 degrees to the header.

 f. Repeating the previous steps is necessary to set each stringer in place.

 g. Last, all treads are installed with galvanized nails.

Figure 16-4 Installing a stringer and leveling with the pier height.

STAIRS WITH 2 × 4 CARRIAGES AND METAL TREAD SUPPORTS

It may be desirable in some situations to construct a set of exterior stairs using metal tread supports. If so, a 2 × 4 would suffice as a carriage onto which the metal tread supports would be lag bolted as in Figure 16-5.

Metal treads can either be ordered from a local metal shop or fabricated on site. In either event the rise and run must be identified so that each tread support is bent to the proper angle and cut for the proper length. Then too, small flanges must be made onto each end with holes bored for passage of the lag bolts. This detail is shown in Figure 16-5. Also, holes must be bored into the tread support surfaces to allow passage of the carriage bolts used to hold the treads in place.

When using this carriage arrangement the pier should be used as explained in the previous section and an angle bracket could be used for securing the carriage to the headers of the deck framing.

BUILT-UP STRINGERS LAID HORIZONTALLY

The standard form of stairs does not always need to be used from deck to ground. Often a set of stairs can be a transition from deck to ground. This means that the stairs can be used as semidecks if each tread depth is made wide enough.

Figure 16-6 shows an example of this idea. Here, two tread surfaces are made 24 in. wide and act as semidecking surfaces. The standard inclined stringer would not be appropriate to use for tread support. Rather a horizontal stringer is more appropriate.

As Figure 16-6 shows, 2 × 4s and 4 × 4s are used in built-up fashion to create the tread-bearing surfaces and rises. Blocks are installed under the lower 4 × 4 to provide sound support as well as elevation above ground level. Construction using this arrangement would be quite simple because the lower 4 × 4 could be leveled; then the short section of 2 × 4 could be nailed to the 4 × 4, and finally the short 4 × 4 could be toenailed to the 2 × 4. Once both or all stringer assemblies are in place, 2 × 4 treads cut to length are nailed to the stringers with 12d galvanized nails.

An alternative design to the horizontal stringer is shown in Figure 16-6a where a 2 × 10 is laid out. A single member of this kind is easily prepared since all cuts are parallel and at right angles.

Frontal view

Metal tread supports

Tread

Metal tread supports

Level used to aid in setting pier block

Pier block

Carriage bolts

Tread

Lag bolts

Carriage

Brace plate

Decking

Figure 16-5 2 x 4 carriage and metal step supports.

Figure 16-6b shows the frontal view of the stairs. The ends of the
4 × 4s show through, but this is seldom a problem since end wood
is expected to show on any open stringers. The openness, which
includes spaces between tread pieces, no riser boards, and slight
elevation above ground level provide complete air circulation and
therefore should provide lasting quality. However, it still is a good
idea to use treated lumber for the exterior stairs.

Figure 16-6 Horizontal stringers for deck-style steps.

PROBLEMS AND QUESTIONS

1. If fungi is to cause rot and decay in wood what percentage of moisture content must be present?
   a. 10%   b. 15%   c. 20%

2. Which type of wood from a tree is best for exterior stringers?
   a. heartwood   b. sapwood

3. Which species are classed as highly resistant to rot and termites?

4. Which type of wood preservative cannot be painted?

5. What might be a problem if brushing on a preservative is the treatment used on a stringer?

6. How can proper drainage of surface moisture extend the life of exterior stairs?

7. Assume that a deck is 48 in. above ground level and stairs are to be built to the ground; what would the average rise be if the maximum rise is 7-1/2 in.?
   a. Six 7-1/2-in. risers   b. Seven 6-13/16-in. F risers

8. When the cut stringer is temporarily set in place how is adjustment for a level tread surface made?

9. What is the purpose of a drift pin in the pier?

10. What size carriage should be used for exterior stairs that are installed with metal tread supports?

11. Cite an example where the stringers would be made from 2 x 12s laid horizontally.

Design Values
for Joists and Rafters

(Courtesy of National Forest Products Association)

DESIGN VALUES FOR JOISTS AND RAFTERS—VISUAL GRADING

These "F_b" values are for use where repetitive members are spaced not more than 24 inches. For wider spacing, the "F_b" values should be reduced 13 percent.
Values for surfaced dry or surfaced green lumber apply at 19 percent maximum moisture content in use.

Species and Grade	Size	Design Value in Bending "F_b"			Modulus of Elasticity "E"	Grading Rules Agency
		Normal Duration	Snow Loading	7-Day Loading		
DOUGLAS FIR—LARCH (Surfaced dry or surfaced green)						
Dense Select Structural		2800	3220	3500	1,900,000	
Select Structural		2400	2760	3000	1,800,000	
Dense No. 1		2400	2760	3000	1,900,000	
No. 1 & Appearance		2050	2360	2560	1,800,000	
	2x4					
Dense No. 2		1950	2240	2440	1,700,000	
No. 2		1650	1900	2060	1,700,000	Western Wood
No. 3		925	1060	1160	1,500,000	Products
Stud		925	1060	1160	1,500,000	Association
						(See notes 1
Construction		1200	1380	1500	1,500,000	and 3)
Standard	2x4	675	780	840	1,500,000	
Utility		325	370	410	1,500,000	
						West Coast
Dense Select Structural		2400	2760	3000	1,900,000	Lumber
Select Structural		2050	2360	2560	1,800,000	Inspection
Dense No. 1	2x5	2050	2360	2560	1,900,000	Bureau
No. 1 & Appearance	and	1750	2010	2190	1,800,000	
	wider					
Dense No. 2		1700	1960	2120	1,700,000	
No. 2		1450	1670	1810	1,700,000	
No. 3		850	980	1060	1,500,000	
Stud		850	980	1060	1,500,000	
DOUGLAS FIR—LARCH (NORTH) (Surfaced dry or surfaced green)						
Dense Select Structural		2800	3220	3500	1,900,000	
Select Structural		2400	2760	3000	1,800,000	
Dense No. 1		2400	2760	3000	1,900,000	
No. 1 & Appearance		2050	2360	2560	1,800,000	
	2x4					
Dense No. 2		1950	2240	2440	1,700,000	Nat'l. Lumber
No. 2		1650	1900	2060	1,700,000	Grades Auth.
No. 3		925	1060	1160	1,500,000	(A Canadian
Stud		925	1060	1160	1,500,000	Agency—
						See notes 1, 2
Construction		1200	1380	1500	1,500,000	and 3)
Standard	2x4	675	780	840	1,500,000	
Utility		325	370	410	1,500,000	
Dense Select Structural		2400	2760	3000	1,900,000	
Select Structural		2050	2360	2560	1,800,000	
Dense No. 1	2x5	2050	2360	2560	1,900,000	
No. 1 & Appearance	and	1750	2010	2190	1,800,000	
	wider					
Dense No. 2		1700	1960	2120	1,700,000	
No. 2		1450	1670	1810	1,700,000	
No. 3		850	980	1060	1,500,000	
Stud		850	980	1060	1,500,000	
DOUGLAS FIR SOUTH (Surfaced dry or surfaced green)						
Select Structural		2300	2640	2880	1,400,000	
No. 1 & Appearance		1950	2240	2440	1,400,000	
No. 2		1600	1840	2000	1,300,000	Western Wood
No. 3	2x4	875	1010	1090	1,100,000	Products
Stud		875	1010	1090	1.100.000	Association
						(See notes 1
Construction		1150	1320	1440	1,100,000	and 3)
Standard	2x4	650	750	810	1,100,000	
Utility		300	340	380	1,100,000	
Select Structural		1950	2240	2440	1,400,000	
No. 1 & Appearance	2x5	1650	1900	2060	1,400,000	
No. 2	and	1350	1550	1690	1,300,000	
No. 3	wider	800	920	1000	1,100,000	
Stud		800	920	1000	1,100,000	

DESIGN VALUES FOR JOISTS AND RAFTERS—VISUAL GRADING (Cont)

These "F_b" values are for use where repetitive members are spaced not more than 24 inches. For wider spacing, the "F_b" values.should be reduced 13 percent.

Values for surfaced dry or surfaced green lumber apply at 19 percent maximum moisture content in use.

Species and Grade	Size	Design Value in Bending "F_b"			Modulus of Elasticity "E"	Grading Rules Agency
		Normal Duration	Snow Loading	7-Day Loading		
EASTERN HEMLOCK—TAMARACK (Surfaced dry or surfaced green)						
Select Structural		2050	2360	2560	1,300,000	Northeastern
No. 1		1750	2010	2190	1,300,000	Lumber
No. 2	2x4	1450	1670	1810	1,100,000	Manufacturers
No. 3		800	920	1000	1,000,000	Association
Appearance		1500	1720	1880	1,300,000	
Stud		800	920	1000	1,000,000	Northern
						Hardwood
Construction		1050	1210	1310	1,000,000	& Pine
Standard	2x4	575	660	720	1,000,000	Manufacturers
Utility		275	320	340	1,000,000	Association
Select Structural		1750	2010	2190	1,300,000	
No. 1 & Appearance	2x5	1500	1720	1880	1,300,000	(See notes 1
No. 2	and	1200	1380	1500	1,100,000	and 3)
No. 3	wider	725	830	910	1,000,000	
Stud		725	830	910	1,000,000	
EASTERN HEMLOCK—TAMARACK (NORTH) (Surfaced dry or surfaced green)						
Select Structural		2050	2360	2560	1,300,000	
No. 1		1750	2010	2190	1,300,000	
No. 2	2x4	1450	1670	1810	1,100,000	Nat'l. Lumber
No. 3		800	920	1000	1,000,000	Grades Auth.
Appearance		1500	1720	1880	1,300,000	(A Canadian
Stud		800	920	1000	1,000,000	Agency—
						See notes 1, 2
Construction		1050	1210	1310	1,000,000	and 3)
Standard	2x4	575	660	720	1,000,000	
Utility		275	320	340	1,000,000	
Select Structural		1750	2010	2190	1,300,000	
No. 1 & Appearance	2x5	1500	1720	1880	1,300,000	
No. 2	and	1200	1380	1500	1,100,000	
No. 3	wider	725	830	910	1,000,000	
Stud		725	830	910	1,000,000	
EASTERN SPRUCE (Surfaced dry or surfaced green)						
Select Structural		1750	2010	2190	1,400,000	Northeastern
No. 1		1500	1720	1880	1,400,000	Lumber
No. 2	2x4	1200	1380	1500	1,200,000	Manufacturers
No. 3		675	780	840	1,100,000	Association
Appearance		1250	1440	1560	1,400,000	
Stud		675	780	840	1,100,000	Northern
						Hardwood
Construction		875	1010	1090	1,100,000	& Pine
Standard	2x4	500	580	620	1,100,000	Manufacturers
Utility		225	260	280	1,100,000	Association
Select Structural		1500	1720	1880	1,400,000	
No. 1 & Appearance	2x5	1250	1440	1560	1,400,000	(See notes 1
No. 2	and	1000	1150	1250	1,200,000	and 3)
No. 3	wider	600	690	750	1,100,000	
Stud		600	690	750	1,100,000	

DESIGN VALUES FOR JOISTS AND RAFTERS—VISUAL GRADING (Cont)

These "F_b" values are for use where repetitive members are spaced not more than 24 inches. For wider spacing, the "F_b" values should be reduced 13 percent.

Values for surfaced dry or surfaced green lumber apply at 19 percent maximum moisture content in use.

Species and Grade	Size	Design Value in Bending "F_b"			Modulus of Elasticity "E"	Grading Rules Agency
		Normal Duration	Snow Loading	7-Day Loading		
IDAHO WHITE PINE (Surfaced dry or surfaced green)						
Select Structural		1550	1780	1940	1,400,000	
No. 1 & Appearance		1300	1500	1620	1,400,000	
No. 2	2x4	1050	1210	1310	1,300,000	Western Wood
No. 3		600	690	750	1,200,000	Products
Stud		600	690	750	1,200,000	Association
Construction		775	890	970	1,200,000	(See notes 1 and 3)
Standard	2x4	425	490	530	1,200,000	
Utility		200	230	250	1,200,000	
Select Structural		1300	1500	1620	1,400,000	
No. 1 & Appearance	2x5	1100	1260	1380	1,400,000	
No. 2	and	925	1060	1160	1,300,000	
No. 3	wider	550	630	690	1,200,000	
Stud		550	630	690	1,200,000	
LODGEPOLE PINE (Surfaced dry or surfaced green)						
Select Structural		1750	2010	2190	1,300,000	
No. 1 & Appearance		1500	1720	1880	1,300,000	
No. 2	2x4	1200	1380	1500	1,200,000	Western Wood
No. 3		675	780	840	1,000,000	Products
Stud		675	780	840	1,000,000	Association
Construction		875	1010	1090	1,000,000	(See notes 1 and 3)
Standard	2x4	500	580	620	1,000,000	
Utility		225	260	280	1,000,000	
Select Structural		1500	1720	1880	1,300,000	
No. 1 & Appearance	2x5	1300	1500	1620	1,300,000	
No. 2	and	1050	1210	1310	1,200,000	
No. 3	wider	625	720	780	1,000,000	
Stud		625	720	780	1,000,000	
MOUNTAIN HEMLOCK (Surfaced dry or surfaced green)						
Select Structural		2000	2300	2500	1,300,000	
No. 1 & Appearance		1700	1960	2120	1,300,000	
No. 2		1400	1610	1750	1,100,000	Western Wood
No. 3	2x4	775	890	970	1,000,000	Products
Stud		775	890	970	1,000,000	Association
Construction		1000	1150	1250	1,000,000	(See notes 1 and 3)
Standard	2x4	575	660	720	1,000,000	
Utility		275	320	340	1,000,000	West Coast
Select Structural		1700	1960	2120	1,300,000	Lumber Inspection
No. 1 & Appearance	2x5	1450	1670	1810	1,300,000	Bureau
No. 2	and	1200	1380	1500	1,100,000	
No. 3	wider	700	800	880	1,000,000	
Stud		700	800	880	1,000,000	

DESIGN VALUES FOR JOISTS AND RAFTERS—VISUAL GRADING (Cont)

These "F_b" values are for use where repetitive members are spaced not more than 24 inches. For wider spacing, the "F_b" values should be reduced 13 percent.

Values for surfaced dry or surfaced green lumber apply at 19 percent maximum moisture content in use.

Species and Grade	Size	Design Value in Bending "F_b"			Modulus of Elasticity "E"	Grading Rules Agency
		Normal Duration	Snow Loading	7-Day Loading		
SOUTHERN PINE (Surfaced dry)						
Select Structural		2300	2640	2880	1,700,000	
Dense Select Structural		2700	3100	3380	1,800,000	
No. 1		1950	2240	2440	1,700,000	
No. 1 Dense		2300	2640	2880	1,800,000	
No. 2	2x4	1650	1900	2060	1,600,000	
No. 2 Dense		1900	2180	2380	1,600,000	
No. 3		900	1040	1120	1,400,000	
No. 3 Dense		1050	1210	1310	1,500,000	Southern
Stud		900	1040	1120	1,400,000	Pine
						Inspection
Construction		1150	1320	1440	1,400,000	Bureau
Standard	2x4	675	780	840	1,400,000	
Utility		300	340	380	1,400,000	(See note 3)
Select Structural		2000	2300	2500	1,700,000	
Dense Select Structural		2350	2700	2940	1,800,000	
No. 1		1700	1960	2120	1,700,000	
No. 1 Dense	2x5	2000	2300	2500	1,800,000	
No. 2	and	1400	1610	1750	1,600,000	
No. 2 Dense	wider	1650	1900	2060	1,600,000	
No. 3		800	920	1000	1,400,000	
No. 3 Dense		925	1060	1160	1,500,000	
Stud		850	980	1060	1,400,000	
SOUTHERN PINE (Surfaced at 15 percent moisture content-KD)						
Select Structural		2500	2880	3120	1,800,000	
Dense Select Structural		2900	3340	3620	1,900,000	
No. 1		2100	2420	2620	1,800,000	
No. 1 Dense		2450	2820	3060	1,900,000	
No. 2	2x4	1750	2010	2190	1,600,000	
No. 2 Dense		2050	2360	2560	1,700,000	Southern
No. 3		975	1120	1220	1,500,000	Pine
No. 3 Dense		1150	1320	1440	1,500,000	Inspection
Stud		975	1120	1220	1,500,000	Bureau
Construction		1250	1440	1560	1,500,000	
Standard	2x4	725	830	910	1,500,000	
Utility		300	340	380	1,500,000	(See note 3)
Select Structural		2150	2470	2690	1,800,000	
Dense Select Structural		2500	2880	3120	1,900,000	
No. 1		1850	2130	2310	1,800,000	
No. 1 Dense	2x5	2150	2470	2690	1,900,000	
	and					
No. 2	wider	1500	1720	1880	1,600,000	
No. 2 Dense		1750	2010	2190	1,700,000	
No. 3		875	1010	1090	1,500,000	
No. 3 Dense		1000	1150	1250	1,500,000	
Stud		900	1040	1120	1,500,000	

DESIGN VALUES FOR JOISTS AND RAFTERS—VISUAL GRADING (Cont)

These "F_b" values are for use where repetitive members are spaced not more than 24 inches. For wider spacing, the "F_b" values should be reduced 13 percent.

Values for surfaced dry or surfaced green lumber apply at 19 percent maximum moisture content in use.

Species and Grade	Size	Design Value in Bending "F_b"			Modulus of Elasticity "E"	Grading Rules Agency
		Normal Duration	Snow Loading	7-Day Loading		
WESTERN HEMLOCK (Surfaced dry or surfaced green)						
Select Structural		2100	2420	2620	1,600,000	Western Wood
No. 1 & Appearance		1800	2070	2250	1,600,000	Products
No. 2	2x4	1450	1670	1810	1,400,000	Association
No. 3		800	920	1000	1,300,000	(See notes 1
Stud		800	920	1000	1,300,000	and 3)
Construction		1050	1210	1310	1,300,000	West Coast
Standard	2x4	600	690	750	1,300,000	Lumber
Utility		275	320	340	1,300,000	Inspection Bureau
Select Structural		1800	2070	2250	1,600,000	
No. 1 & Appearance	2x5	1550	1780	1940	1,600,000	
No. 2	and	1250	1440	1560	1,400,000	
No. 3	wider	750	860	940	1,300,000	
Stud		750	860	940	1,300,000	
WESTERN WHITE PINE (Surfaced dry or surfaced green)						
Select Structural		1550	1780	1940	1,400,000	
No. 1 & Appearance		1300	1500	1620	1,400,000	
No. 2	2x4	1050	1210	1310	1,300,000	
No. 3		600	690	750	1,200,000	
Stud		600	690	750	1,200,000	Nat'l. Lumber
Construction		775	890	970	1,200,000	Grades Auth.
Standard	2x4	425	490	530	1,200,000	(A Canadian
Utility		200	230	250	1,200,000	Agency—
Select Structural		1300	1500	1620	1,400,000	See notes 1, 2
No. 1 & Appearance	2x5	1150	1320	1440	1,400,000	and 3)
No. 2	and	925	1060	1160	1,300,000	
No. 3	wider	550	630	690	1,200,000	
Stud		550	630	690	1,200,000	
WHITE WOODS (WESTERN WOODS) (Surfaced dry or surfaced green)						
Select Structural		1550	1780	1940	1,100,000	
No. 1 & Appearance		1300	1500	1620	1,100,000	
No. 2	2x4	1050	1210	1310	1,000,000	
No. 3		600	690	750	900,000	
Stud		600	690	750	900,000	Western Wood
Construction		775	890	970	900,000	Products
Standard	2x4	425	490	530	900,000	Association
Utility		200	230	250	900,000	(See notes 1 and 3)
Select Structural		1300	1500	1620	1,100,000	
No. 1 & Appearance	2x5	1100	1260	1380	1,100,000	
No. 2	and	925	1060	1160	1,000,000	
No. 3	wider	550	630	690	900,000	
Stud		550	630	690	900,000	

Table W-1 Footnotes Applicable to Visually Graded Joists and Rafters

1. When 2" lumber is manufactured at a maximum moisture content of 15 percent (grade-marked MC-15) and used in a condition where the moisture content does not exceed 15 percent the design values shown in Table W-1 for "surfaced dry or surfaced green" lumber may be increased eight percent (8%) for design value in bending "F_b", and five percent (5%) for Modulus of Elasticity "E".

2. National Lumber Grades Authority is the Canadian rules writing agency responsible for preparation, maintenance and dissemination of a uniform softwood lumber grading rule for all Canadian species.

3. Design values for Stud grade in 2x5 and wider size classifications apply to 5" and 6" widths only.

Span Tables
for Joists and Rafters

(Courtesy of National Forest Products Association)

FLOOR JOISTS
40 Lbs. Per Sq. Ft. Live Load
(All rooms except those used for sleeping areas and attic floors.)

JOIST SIZE (IN)	SPACING (IN)	Modulus of Elasticity, "E", in 1,000,000 psi								
		0.4	0.5	0.6	0.7	0.8	0.9	1.0	1.1	1.2
2x6	12.0	6-9 / 450	7-3 / 520	7-9 / 590	8-2 / 660	8-6 / 720	8-10 / 780	9-2 / 830	9-6 / 890	9-9 / 940
	13.7	6-6 / 470	7-0 / 550	7-5 / 620	7-9 / 690	8-2 / 750	8-6 / 810	8-9 / 870	9-1 / 930	9-4 / 980
	16.0	6-2 / 500	6-7 / 580	7-0 / 650	7-5 / 720	7-9 / 790	8-0 / 860	8-4 / 920	8-7 / 980	8-10 / 1040
	19.2	5-9 / 530	6-3 / 610	6-7 / 690	7-0 / 770	7-3 / 840	7-7 / 910	7-10 / 970	8-1 / 1040	8-4 / 1100
	24.0	5-4 / 570	5-9 / 660	6-2 / 750	6-6 / 830	6-9 / 900	7-0 / 980	7-3 / 1050	7-6 / 1120	7-9 / 1190
	32.0					6-2 / 1010	6-5 / 1090	6-7 / 1150	6-10 / 1230	7-0 / 1300
2x8	12.0	8-11 / 450	9-7 / 520	10-2 / 590	10-9 / 660	11-3 / 720	11-8 / 780	12-1 / 830	12-6 / 890	12-10 / 940
	13.7	8-6 / 470	9-2 / 550	9-9 / 620	10-3 / 690	10-9 / 750	11-2 / 810	11-7 / 870	11-11 / 930	12-3 / 980
	16.0	8-1 / 500	8-9 / 580	9-3 / 650	9-9 / 720	10-2 / 790	10-7 / 850	11-0 / 920	11-4 / 980	11-8 / 1040
	19.2	7-7 / 530	8-2 / 610	8-9 / 690	9-2 / 770	9-7 / 840	10-0 / 910	10-4 / 970	10-8 / 1040	11-0 / 1100
	24.0	7-1 / 570	7-7 / 660	8-1 / 750	8-6 / 830	8-11 / 900	9-3 / 980	9-7 / 1050	9-11 / 1120	10-2 / 1190
	32.0					8-1 / 990	8-5 / 1080	8-9 / 1170	9-0 / 1230	9-3 / 1300
2x10	12.0	11-4 / 450	12-3 / 520	13-0 / 590	13-8 / 660	14-4 / 720	14-11 / 780	15-5 / 830	15-11 / 890	16-5 / 940
	13.7	10-10 / 470	11-8 / 550	12-5 / 620	13-1 / 690	13-8 / 750	14-3 / 810	14-9 / 870	15-3 / 930	15-8 / 980
	16.0	10-4 / 500	11-1 / 580	11-10 / 650	12-5 / 720	13-0 / 790	13-6 / 850	14-0 / 920	14-6 / 980	14-11 / 1040
	19.2	9-9 / 530	10-6 / 610	11-1 / 690	11-8 / 770	12-3 / 840	12-9 / 910	13-2 / 970	13-7 / 1040	14-0 / 1100
	24.0	9-0 / 570	9-9 / 660	10-4 / 750	10-10 / 830	11-4 / 900	11-10 / 980	12-3 / 1050	12-8 / 1120	13-0 / 1190
	32.0					10-4 / 1000	10-9 / 1080	11-1 / 1150	11-6 / 1240	11-10 / 1310
2x12	12.0	13-10 / 450	14-11 / 520	15-10 / 590	16-8 / 660	17-5 / 720	18-1 / 780	18-9 / 830	19-4 / 890	19-11 / 940
	13.7	13-3 / 470	14-3 / 550	15-2 / 620	15-11 / 690	16-8 / 750	17-4 / 810	17-11 / 870	18-6 / 930	19-1 / 980
	16.0	12-7 / 500	13-6 / 580	14-4 / 650	15-2 / 720	15-10 / 790	16-5 / 860	17-0 / 920	17-7 / 980	18-1 / 1040
	19.2	11-10 / 530	12-9 / 610	13-6 / 690	14-3 / 770	14-11 / 840	15-6 / 910	16-0 / 970	16-7 / 1040	17-0 / 1100
	24.0	11-0 / 570	11-10 / 660	12-7 / 750	13-3 / 830	13-10 / 900	14-4 / 980	14-11 / 1050	15-4 / 1120	15-10 / 1190
	32.0					12-7 / 1000	13-1 / 1080	13-6 / 1150	13-11 / 1220	14-4 / 1300

Note: The required extreme fiber stress in bending, "F_b", in pounds per square inch is shown below each span.

DESIGN CRITERIA:
Deflection - For 40 lbs. per sq. ft. live load.
  Limited to span in inches divided by 360.
Strength - Live Load of 40 lbs. per sq. ft. plus
  dead load of 10 lbs. per sq. ft. determines the
  required fiber stress value.

1.3	1.4	1.5	1.6	1.7	1.8	1.9	2.0	2.2	2.4
10-0 / 990	10-3 / 1040	10-6 / 1090	10-9 / 1140	10-11 / 1190	11-2 / 1230	11-4 / 1280	11-7 / 1320	11-11 / 1410	12-3 / 1490
9-7 / 1040	9-10 / 1090	10-0 / 1140	10-3 / 1190	10-6 / 1240	10-8 / 1290	10-10 / 1340	11-1 / 1380	11-5 / 1470	11-9 / 1560
9-1 / 1090	9-4 / 1150	9-6 / 1200	9-9 / 1250	9-11 / 1310	10-2 / 1360	10-4 / 1410	10-6 / 1460	10-10 / 1550	11-2 / 1640
8-7 / 1160	8-9 / 1220	9-0 / 1280	9-2 / 1330	9-4 / 1390	9-6 / 1440	9-8 / 1500	9-10 / 1550	10-2 / 1650	10-6 / 1750
7-11 / 1250	8-2 / 1310	8-4 / 1380	8-6 / 1440	8-8 / 1500	8-10 / 1550	9-0 / 1610	9-2 / 1670	9-6 / 1780	9-9 / 1880
7-3 / 1390	7-5 / 1450	7-7 / 1520	7-9 / 1590	7-11 / 1660	8-0 / 1690	8-2 / 1760	8-4 / 1840	8-7 / 1950	8-10 / 2060
13-2 / 990	13-6 / 1040	13-10 / 1090	14-2 / 1140	14-5 / 1190	14-8 / 1230	15-0 / 1280	15-3 / 1320	15-9 / 1410	16-2 / 1490
12-7 / 1040	12-11 / 1090	13-3 / 1140	13-6 / 1190	13-10 / 1240	14-1 / 1290	14-4 / 1340	14-7 / 1380	15-0 / 1470	15-6 / 1560
12-0 / 1090	12-3 / 1150	12-7 / 1200	12-10 / 1250	13-1 / 1310	13-4 / 1360	13-7 / 1410	13-10 / 1460	14-3 / 1550	14-8 / 1640
11-3 / 1160	11-7 / 1220	11-10 / 1280	12-1 / 1330	12-4 / 1390	12-7 / 1440	12-10 / 1500	13-0 / 1550	13-5 / 1650	13-10 / 1750
10-6 / 1250	10-9 / 1310	11-0 / 1380	11-3 / 1440	11-5 / 1500	11-8 / 1550	11-11 / 1610	12-1 / 1670	12-6 / 1780	12-10 / 1880
9-6 / 1370	9-9 / 1450	10-0 / 1520	10-2 / 1570	10-5 / 1650	10-7 / 1700	10-10 / 1790	11-0 / 1840	11-4 / 1950	11-8 / 2070
16-10 / 990	17-3 / 1040	17-8 / 1090	18-0 / 1140	18-5 / 1190	18-9 / 1230	19-1 / 1280	19-5 / 1320	20-1 / 1410	20-8 / 1490
16-1 / 1040	16-6 / 1090	16-11 / 1140	17-3 / 1190	17-7 / 1240	17-11 / 1290	18-3 / 1340	18-7 / 1380	19-2 / 1470	19-9 / 1560
15-3 / 1090	15-8 / 1150	16-0 / 1200	16-5 / 1250	16-9 / 1310	17-0 / 1360	17-4 / 1410	17-8 / 1460	18-3 / 1550	18-9 / 1640
14-5 / 1160	14-9 / 1220	15-1 / 1280	15-5 / 1330	15-9 / 1390	16-0 / 1440	16-4 / 1500	16-7 / 1550	17-2 / 1650	17-8 / 1750
13-4 / 1250	13-8 / 1310	14-0 / 1380	14-4 / 1440	14-7 / 1500	14-11 / 1550	15-2 / 1610	15-5 / 1670	15-11 / 1780	16-5 / 1880
12-2 / 1380	12-5 / 1440	12-9 / 1520	13-0 / 1580	13-3 / 1640	13-6 / 1700	13-9 / 1770	14-0 / 1830	14-6 / 1970	14-11 / 2080
20-6 / 990	21-0 / 1040	21-6 / 1090	21-11 / 1140	22-5 / 1190	22-10 / 1230	23-3 / 1280	23-7 / 1320	24-5 / 1410	25-1 / 1490
19-7 / 1040	20-1 / 1090	20-6 / 1140	21-0 / 1190	21-5 / 1240	21-10 / 1290	22-3 / 1340	22-7 / 1380	23-4 / 1470	24-0 / 1560
18-7 / 1090	19-1 / 1150	19-6 / 1200	19-11 / 1250	20-4 / 1310	20-9 / 1360	21-1 / 1410	21-6 / 1460	22-2 / 1550	22-10 / 1640
17-6 / 1160	17-11 / 1220	18-4 / 1280	18-9 / 1330	19-2 / 1390	19-6 / 1440	19-10 / 1500	20-2 / 1550	20-10 / 1650	21-6 / 1750
16-3 / 1250	16-8 / 1310	17-0 / 1380	17-5 / 1440	17-9 / 1500	18-1 / 1550	18-5 / 1610	18-9 / 1670	19-4 / 1780	19-11 / 1880
14-9 / 1380	15-2 / 1450	15-6 / 1520	15-10 / 1580	16-2 / 1650	16-5 / 1700	16-9 / 1770	17-0 / 1830	17-7 / 1950	18-1 / 2070

FLOOR JOISTS
30 lbs. Per Sq. Ft. Live Load
(All rooms used for sleeping areas and attic floors.)

JOIST SIZE (IN)	SPACING (IN)	Modulus of Elasticity, "E", in 1,000,000 psi								
		0.4	0.5	0.6	0.7	0.8	0.9	1.0	1.1	1.2
2x6	12.0	7-5 440	8-0 510	8-6 570	8-11 640	9-4 700	9-9 750	10-1 810	10-5 860	10-9 910
	13.7	7-1 460	7-8 530	8-2 600	8-7 670	8-11 730	9-4 790	9-8 840	10-0 900	10-3 950
	16.0	6-9 480	7-3 560	7-9 630	8-2 700	8-6 770	8-10 830	9-2 890	9-6 950	9-9 1000
	19.2	6-4 510	6-10 600	7-3 670	7-8 740	8-0 810	8-4 880	8-8 940	8-11 1010	9-2 1070
	24.0	5-11 550	6-4 640	6-9 720	7-1 800	7-5 880	7-9 950	8-0 1020	8-3 1080	8-6 1150
	32.0					6-9 960	7-0 1040	7-3 1110	7-6 1190	7-9 1270
2x8	12.0	9-10 440	10-7 510	11-3 570	11-10 640	12-4 700	12-10 750	13-4 810	13-9 860	14-2 910
	13.7	9-4 460	10-1 530	10-9 600	11-4 670	11-10 730	12-3 790	12-9 840	13-2 900	13-6 950
	16.0	8-11 480	9-7 560	10-2 630	10-9 700	11-3 770	11-8 830	12-1 890	12-6 950	12-10 1000
	19.2	8-5 510	9-0 600	9-7 670	10-1 740	10-7 810	11-0 880	11-4 940	11-9 1010	12-1 1070
	24.0	7-9 550	8-5 640	8-11 720	9-4 800	9-10 880	10-2 950	10-7 1020	10-11 1080	11-3 1150
	32.0					8-11 970	9-3 1040	9-7 1120	9-11 1200	10-2 1260
2x10	12.0	12-6 440	13-6 510	14-4 570	15-1 640	15-9 700	16-5 750	17-0 810	17-6 860	18-0 910
	13.7	11-11 460	12-11 530	13-8 600	14-5 670	15-1 730	15-8 790	16-3 840	16-9 900	17-3 950
	16.0	11-4 480	12-3 560	13-0 630	13-8 700	14-4 770	14-11 830	15-5 890	15-11 950	16-5 1000
	19.2	10-8 510	11-6 600	12-3 670	12-11 740	13-6 810	14-0 880	14-6 940	15-0 1010	15-5 1070
	24.0	9-11 550	10-8 640	11-4 720	11-11 800	12-6 880	13-0 950	13-6 1020	13-11 1080	14-4 1150
	32.0					11-4 960	11-10 1050	12-3 1120	12-8 1200	13-0 1260
2x12	12.0	15-2 440	16-5 510	17-5 570	18-4 640	19-2 700	19-11 750	20-8 810	21-4 860	21-11 910
	13.7	14-7 460	15-8 530	16-8 600	17-6 670	18-4 730	19-1 790	19-9 840	20-5 900	21-0 950
	16.0	13-10 480	14-11 560	15-10 630	16-8 700	17-5 770	18-1 830	18-9 890	19-4 950	19-11 1000
	19.2	13-0 510	14-0 600	14-11 670	15-8 740	16-5 810	17-0 880	17-8 940	18-3 1010	18-9 1070
	24.0	12-1 550	13-0 640	13-10 720	14-7 800	15-2 880	15-10 950	16-5 1020	16-11 1080	17-5 1150
	32.0					13-10 970	14-4 1040	14-11 1130	15-4 1190	15-10 1270

Note: The required extreme fiber stress in bending, "F_b", in pounds per square inch is shown below each span.

DESIGN CRITERIA:
Deflection · For 30 lbs. per sq. ft. live load.
  Limited to span in inches divided by 360.
Strength · Live Load of 30 lbs. per sq. ft. plus
  dead load of 10 lbs. per sq. ft. determines
  the required fiber stress value.

1.3	1.4	1.5	1.6	1.7	1.8	1.9	2.0	2.2	2.4
11-0 960	11-3 1010	11-7 1060	11-10 1100	12-0 1150	12-3 1200	12-6 1240	12-9 1280	13-1 1370	13-6 1450
10-6 1010	10-10 1060	11-1 1110	11-3 1160	11-6 1200	11-9 1250	11-11 1300	12-2 1340	12-7 1430	12-11 1510
10-0 1060	10-3 1110	10-6 1160	10-9 1220	10-11 1270	11-2 1320	11-4 1360	11-7 1410	11-11 1500	12-3 1590
9-5 1130	9-8 1180	9-10 1240	10-1 1290	10-4 1350	10-6 1400	10-8 1450	10-10 1500	11-3 1600	11-7 1690
8-9 1210	8-11 1270	9-2 1330	9-4 1390	9-7 1450	9-9 1510	9-11 1560	10-1 1620	10-5 1720	10-9 1820
7-11 1330	8-2 1410	8-4 1470	8-6 1530	8-8 1590	8-10 1650	9-0 1710	9-2 1780	9-6 1910	9-9 2010
14-6 960	14-11 1010	15-3 1060	15-7 1100	15-10 1150	16-2 1200	16-6 1240	16-9 1280	17-4 1370	17-10 1450
13-11 1010	14-3 1060	14-7 1110	14-11 1160	15-2 1200	15-6 1250	15-9 1300	16-0 1340	16-7 1430	17-0 1510
13-2 1060	13-6 1110	13-10 1160	14-2 1220	14-5 1270	14-8 1320	15-0 1360	15-3 1410	15-9 1500	16-2 1590
12-5 1130	12-9 1180	13-0 1240	13-4 1290	13-7 1350	13-10 1400	14-1 1450	14-4 1500	14-9 1600	15-3 1690
11-6 1210	11-10 1270	12-1 1330	12-4 1390	12-7 1450	12-10 1510	13-1 1560	13-4 1620	13-9 1720	14-2 1820
10-6 1340	10-9 1410	11-0 1470	11-3 1540	11-5 1590	11-8 1660	11-11 1730	12-1 1780	12-6 1900	12-10 2010
18-6 960	19-0 1010	19-5 1060	19-10 1100	20-3 1150	20-8 1200	21-0 1240	21-5 1280	22-1 1370	22-9 1450
17-9 1010	18-2 1060	18-7 1110	19-0 1160	19-4 1200	19-9 1250	20-1 1300	20-5 1340	21-1 1430	21-9 1510
16-10 1060	17-3 1110	17-8 1160	18-0 1220	18-5 1270	18-9 1320	19-1 1360	19-5 1410	20-1 1500	20-8 1590
15-10 1130	16-3 1180	16-7 1240	17-0 1290	17-4 1350	17-8 1400	18-0 1450	18-3 1500	18-10 1600	19-5 1690
14-8 1210	15-1 1270	15-5 1330	15-9 1390	16-1 1450	16-5 1510	16-8 1560	17-0 1620	17-6 1720	18-0 1820
13-4 1330	13-8 1400	14-0 1470	14-4 1540	14-7 1590	14-11 1660	15-2 1720	15-5 1780	15-11 1890	16-5 2020
22-6 960	23-1 1010	23-7 1060	24-2 1100	24-8 1150	25-1 1200	25-7 1240	26-0 1280	26-10 1370	27-8 1450
21-7 1010	22-1 1060	22-7 1110	23-1 1160	23-7 1200	24-0 1250	24-5 1300	24-10 1340	25-8 1430	26-5 1510
20-6 1060	21-0 1110	21-6 1160	21-11 1220	22-5 1270	22-10 1320	23-3 1360	23-7 1410	24-5 1500	25-1 1590
19-3 1130	19-9 1180	20-2 1240	20-8 1290	21-1 1350	21-6 1400	21-10 1450	22-3 1500	22-11 1600	23-7 1690
17-11 1210	18-4 1270	18-9 1330	19-2 1390	19-7 1450	19-11 1510	20-3 1560	20-8 1620	21-4 1720	21-11 1820
16-3 1340	16-8 1400	17-0 1460	17-5 1530	17-9 1590	18-1 1650	18-5 1720	18-9 1780	19-4 1890	19-11 2010

CEILING JOISTS

20 Lbs. Per Sq. Ft. Live Load
(Limited attic storage where development of future rooms is not possible)
(Drywall Ceiling)

JOIST SIZE (IN)	SPACING (IN)	Modulus of Elasticity, "E", in 1,000,000 psi								
		0.4	0.5	0.6	0.7	0.8	0.9	1.0	1.1	1.2
2x4	12.0	6-2 560	6-8 660	7-1 740	7-6 820	7-10 900	8-1 970	8-5 1040	8-8 1110	8-11 1170
	13.7	5-11 590	6-5 690	6-9 770	7-2 860	7-6 940	7-9 1010	8-1 1090	8-4 1160	8-7 1230
	16.0	5-8 620	6-1 720	6-5 810	6-9 900	7-1 990	7-5 1070	7-8 1140	7-11 1220	8-1 1290
	19.2	5-4 660	5-9 770	6-1 870	6-5 960	6-8 1050	6-11 1130	7-2 1220	7-5 1300	7-8 1370
	24.0	4-11 710	5-4 830	5-8 930	5-11 1030	6-2 1130	6-5 1220	6-8 1310	6-11 1400	7-1 1480
2x6	12.0	9-9 560	10-6 660	11-2 740	11-9 820	12-3 900	12-9 970	13-3 1040	13-8 1110	14-1 1170
	13.7	9-4 590	10-0 690	10-8 770	11-3 860	11-9 940	12-3 1010	12-8 1090	13-1 1160	13-5 1230
	16.0	8-10 620	9-6 720	10-2 810	10-8 900	11-2 990	11-7 1070	12-0 1140	12-5 1220	12-9 1290
	19.2	8-4 660	9-0 770	9-6 870	10-0 960	10-6 1050	10-11 1130	11-4 1220	11-8 1300	12-0 1370
	24.0	7-9 710	8-4 830	8-10 930	9-4 1030	9-9 1130	10-2 1220	10-6 1310	10-10 1400	11-2 1480
2x8	12.0	12-10 560	13-10 660	14-8 740	15-6 820	16-2 900	16-10 970	17-5 1040	18-0 1110	18-6 1170
	13.7	12-3 590	13-3 690	14-1 770	14-10 860	15-6 940	16-1 1010	16-8 1090	17-2 1160	17-9 1230
	16.0	11-8 620	12-7 720	13-4 810	14-1 900	14-8 990	15-3 1070	15-10 1140	16-4 1220	16-10 1290
	19.2	11-0 660	11-10 770	12-7 870	13-3 960	13-10 1050	14-5 1130	14-11 1220	15-5 1300	15-10 1370
	24.0	10-2 710	11-0 830	11-8 930	12-3 1030	12-10 1130	13-4 1220	13-10 1310	14-3 1400	14-8 1480
2x10	12.0	16-5 560	17-8 660	18-9 740	19-9 820	20-8 900	21-6 970	22-3 1040	22-11 1110	23-8 1170
	13.7	15-8 590	16-11 690	17-11 770	18-11 860	19-9 940	20-6 1010	21-3 1090	21-11 1160	22-7 1230
	16.0	14-11 620	16-0 720	17-0 810	17-11 900	18-9 990	19-6 1070	20-2 1140	20-10 1220	21-6 1290
	19.2	14-0 660	15-1 770	16-0 870	16-11 960	17-8 1050	18-4 1130	19-0 1220	19-7 1300	20-2 1370
	24.0	13-0 710	14-0 830	14-11 930	15-8 1030	16-5 1130	17-0 1220	17-8 1310	18-3 1400	18-9 1480

Note: The required extreme fiber stress in bending, "F_b", in pounds per square inch is shown below each span.

DESIGN CRITERIA:
Deflection - For 20 lbs. per sq. ft. live load.
 Limited to span in inches divided by 240.
Strength - live load of 20 lbs. per sq. ft. plus dead
 load of 10 lbs. per sq. ft. determines required
 fiber stress value.

1.3	1.4	1.5	1.6	1.7	1.8	1.9	2.0	2.2	2.4
9-2 1240	9-5 1300	9-8 1360	9-10 1420	10-0 1480	10-3 1540	10-5 1600	10-7 1650	10-11 1760	11-3 1860
8-9 1300	9-0 1360	9-3 1420	9-5 1490	9-7 1550	9-9 1610	10-0 1670	10-2 1730	10-6 1840	10-9 1950
8-4 1360	8-7 1430	8-9 1500	8-11 1570	9-1 1630	9-4 1690	9-6 1760	9-8 1820	9-11 1940	10-3 2050
7-10 1450	8-1 1520	8-3 1590	8-5 1660	8-7 1730	8-9 1800	8-11 1870	9-1 1930	9-4 2060	9-8 2180
7-3 1560	7-6 1640	7-8 1720	7-10 1790	8-0 1870	8-1 1940	8-3 2010	8-5 2080	8-8 2220	8-11 2350
14-5 1240	14-9 1300	15-2 1360	15-6 1420	15-9 1480	16-1 1540	16-4 1600	16-8 1650	17-2 1760	17-8 1860
13-10 1300	14-2 1360	14-6 1420	14-9 1490	15-1 1550	15-5 1610	15-8 1670	15-11 1730	16-5 1840	16-11 1950
13-1 1360	13-5 1430	13-9 1500	14-1 1570	14-4 1630	14-7 1690	14-11 1760	15-2 1820	15-7 1940	16-1 2050
12-4 1450	12-8 1520	12-11 1590	13-3 1660	13-6 1730	13-9 1800	14-0 1870	14-3 1930	14-8 2060	15-2 2180
11-5 1560	11-9 1640	12-0 1720	12-3 1790	12-6 1870	12-9 1940	13-0 2010	13-3 2080	13-8 2220	14-1 2350
19-0 1240	19-6 1300	19-11 1360	20-5 1420	20-10 1480	21-2 1540	21-7 1600	21-11 1650	22-8 1760	23-4 1860
18-2 1300	18-8 1360	19-1 1420	19-6 1490	19-11 1550	20-3 1610	20-8 1670	21-0 1730	21-8 1840	22-4 1950
17-3 1360	17-9 1430	18-2 1500	18-6 1570	18-11 1630	19-3 1690	19-7 1760	19-11 1820	20-7 1940	21-2 2050
16-3 1450	16-8 1520	17-1 1590	17-5 1660	17-9 1730	18-2 1800	18-5 1870	18-9 1930	19-5 2060	19-11 2180
15-1 1560	15-6 1640	15-10 1720	16-2 1790	16-6 1870	16-10 1940	17-2 2010	17-5 2080	18-0 2220	18-6 2350
24-3 1240	24-10 1300	25-5 1360	26-0 1420	26-6 1480	27-1 1540	27-6 1600	28-0 1650	28-11 1760	29-9 1860
23-3 1300	23-9 1360	24-4 1420	24-10 1490	25-5 1550	25-10 1610	26-4 1670	26-10 1730	27-8 1840	28-6 1950
22-1 1360	22-7 1430	23-2 1500	23-8 1570	24-1 1630	24-7 1690	25-0 1760	25-5 1820	26-3 1940	27-1 2050
20-9 1450	21-3 1520	21-9 1590	22-3 1660	22-8 1730	23-2 1800	23-7 1870	23-11 1930	24-9 2060	25-5 2180
19-3 1560	19-9 1640	20-2 1720	20-8 1790	21-1 1870	21-6 1940	21-10 2010	22-3 2080	22-11 2220	23-8 2350

LOW SLOPE RAFTERS
Slope 3 in 12 or less - 20 Lbs. Per Sq. Ft. Live Load
(No Finished Ceiling)

DESIGN CRITERIA:
Strength - 10 lbs. per sq. ft. dead load plus 20 lbs. per sq. ft. live load determines required fiber stress.
Deflection - For 20 lbs. per sq. ft. live load. Limited to span in inches divided by 240.

RAFTER SIZE (IN)	SPACING (IN)	Allowable Extreme Fiber Stress in Bending, "F_b" (psi).										
		300	400	500	600	700	800	900	1000	1100	1200	1300
2x6	12.0	7-1 / 0.15	8-2 / 0.24	9-2 / 0.33	10-0 / 0.44	10-10 / 0.55	11-7 / 0.67	12-4 / 0.80	13-0 / 0.94	13-7 / 1.09	14-2 / 1.24	14-9 / 1.40
	13.7	6-8 / 0.14	7-8 / 0.22	8-7 / 0.31	9-5 / 0.41	10-2 / 0.52	10-10 / 0.63	11-6 / 0.75	12-2 / 0.88	12-9 / 1.02	13-3 / 1.16	13-10 / 1.31
	16.0	6-2 / 0.13	7-1 / 0.21	7-11 / 0.29	8-8 / 0.38	9-5 / 0.48	10-0 / 0.58	10-8 / 0.70	11-3 / 0.82	11-9 / 0.94	12-4 / 1.07	12-10 / 1.21
	19.2	5-7 / 0.12	6-6 / 0.19	7-3 / 0.26	7-11 / 0.35	8-7 / 0.44	9-2 / 0.53	9-9 / 0.64	10-3 / 0.75	10-9 / 0.86	11-3 / 0.98	11-8 / 1.10
	24.0	5-0 / 0.11	5-10 / 0.17	6-6 / 0.24	7-1 / 0.31	7-8 / 0.39	8-2 / 0.48	8-8 / 0.57	9-2 / 0.67	9-7 / 0.77	10-0 / 0.88	10-5 / 0.99
2x8	12.0	9-4 / 0.15	10-10 / 0.24	12-1 / 0.33	13-3 / 0.44	14-4 / 0.55	15-3 / 0.67	16-3 / 0.80	17-1 / 0.94	17-11 / 1.09	18-9 / 1.24	19-6 / 1.40
	13.7	8-9 / 0.14	10-1 / 0.22	11-4 / 0.31	12-5 / 0.41	13-4 / 0.52	14-4 / 0.63	15-2 / 0.75	16-0 / 0.88	16-9 / 1.02	17-6 / 1.16	18-3 / 1.31
	16.0	8-1 / 0.13	9-4 / 0.21	10-6 / 0.29	11-6 / 0.38	12-5 / 0.48	13-3 / 0.58	14-0 / 0.70	14-10 / 0.82	15-6 / 0.94	16-3 / 1.07	16-10 / 1.21
	19.2	7-5 / 0.12	8-7 / 0.19	9-7 / 0.26	10-6 / 0.35	11-4 / 0.44	12-1 / 0.53	12-10 / 0.64	13-6 / 0.75	14-2 / 0.86	14-10 / 0.98	15-5 / 1.10
	24.0	6-7 / 0.11	7-8 / 0.17	8-7 / 0.24	9-4 / 0.31	10-1 / 0.39	10-10 / 0.48	11-6 / 0.57	12-1 / 0.67	12-8 / 0.77	13-3 / 0.88	13-9 / 0.99
2x10	12.0	11-11 / 0.15	13-9 / 0.24	15-5 / 0.33	16-11 / 0.44	18-3 / 0.55	19-6 / 0.67	20-8 / 0.80	21-10 / 0.94	22-10 / 1.09	23-11 / 1.24	24-10 / 1.40
	13.7	11-2 / 0.14	12-11 / 0.22	14-5 / 0.31	15-10 / 0.41	17-1 / 0.52	18-3 / 0.63	19-4 / 0.75	20-5 / 0.88	21-5 / 1.02	22-4 / 1.16	23-3 / 1.31
	16.0	10-4 / 0.13	11-11 / 0.21	13-4 / 0.29	14-8 / 0.38	15-10 / 0.48	16-11 / 0.58	17-11 / 0.70	18-11 / 0.82	19-10 / 0.94	20-8 / 1.07	21-6 / 1.21
	19.2	9-5 / 0.12	10-11 / 0.19	12-2 / 0.26	13-4 / 0.35	14-5 / 0.44	15-5 / 0.53	16-4 / 0.64	17-3 / 0.75	18-1 / 0.86	18-11 / 0.98	19-8 / 1.10
	24.0	8-5 / 0.11	9-9 / 0.17	10-11 / 0.24	11-11 / 0.31	12-11 / 0.39	13-9 / 0.48	14-8 / 0.57	15-5 / 0.67	16-2 / 0.77	16-11 / 0.88	17-7 / 0.99
2x12	12.0	14-6 / 0.15	16-9 / 0.24	18-9 / 0.33	20-6 / 0.44	22-2 / 0.55	23-9 / 0.67	25-2 / 0.80	26-6 / 0.94	27-10 / 1.09	29-1 / 1.24	30-3 / 1.40
	13.7	13-7 / 0.14	15-8 / 0.22	17-6 / 0.31	19-3 / 0.41	20-9 / 0.52	22-2 / 0.63	23-6 / 0.75	24-10 / 0.88	26-0 / 1.02	27-2 / 1.16	28-3 / 1.31
	16.0	12-7 / 0.13	14-6 / 0.21	16-3 / 0.29	17-9 / 0.38	19-3 / 0.48	20-6 / 0.58	21-9 / 0.70	23-0 / 0.82	24-1 / 0.94	25-2 / 1.07	26-2 / 1.21
	19.2	11-6 / 0.12	13-3 / 0.19	14-10 / 0.26	16-3 / 0.35	17-6 / 0.44	18-9 / 0.53	19-11 / 0.64	21-0 / 0.75	22-0 / 0.86	23-0 / 0.98	23-11 / 1.10
	24.0	10-3 / 0.11	11-10 / 0.17	13-3 / 0.24	14-6 / 0.31	15-8 / 0.39	16-9 / 0.48	17-9 / 0.57	18-9 / 0.67	19-8 / 0.77	20-6 / 0.88	21-5 / 0.99

Note: The required modulus of elasticity, "E", in 1,000,000 pounds per square inch is shown below each span.

RAFTERS: Spans are measured along the horizontal projection and loads are considered as applied on the horizontal projection.

	Allowable Extreme Fiber Stress in Bending, "F_b" (psi).										RAFTER	SPACING SIZE
1400	1500	1600	1700	1800	1900	2000	2100	2200	2400	(IN)	(IN)
15-4 / 1.56	15-11 / 1.73	16-5 / 1.91	16-11 / 2.09	17-5 / 2.28	17-10 / 2.47					12.0	
14-4 / 1.46	14-10 / 1.62	15-4 / 1.78	15-10 / 1.95	16-3 / 2.13	16-9 / 2.31	17-2 / 2.49				13.7	
13-3 / 1.35	13-9 / 1.50	14-2 / 1.65	14-8 / 1.81	15-1 / 1.97	15-6 / 2.14	15-11 / 2.31	16-3 / 2.48			16.0	2x6
12-2 / 1.23	12-7 / 1.37	13-0 / 1.51	13-4 / 1.65	13-9 / 1.80	14-2 / 1.95	14-6 / 2.11	14-10 / 2.27	15-2 / 2.43		19.2	
10-10 / 1.10	11-3 / 1.22	11-7 / 1.35	11-11 / 1.48	12-4 / 1.61	12-8 / 1.75	13-0 / 1.89	13-3 / 2.03	13-7 / 2.18	14-2 / 2.48	24.0	
20-3 / 1.56	20-11 / 1.73	21-7 / 1.91	22-3 / 2.09	22-11 / 2.28	23-7 / 2.47					12.0	
18-11 / 1.46	19-7 / 1.62	20-3 / 1.78	20-10 / 1.95	21-5 / 2.13	22-0 / 2.31	22-7 / 2.49				13.7	
17-6 / 1.35	18-2 / 1.50	18-9 / 1.65	19-4 / 1.81	19-10 / 1.97	20-5 / 2.14	20-11 / 2.31	21-5 / 2.48			16.0	2x8
16-0 / 1.23	16-7 / 1.37	17-1 / 1.51	17-7 / 1.65	18-2 / 1.80	18-7 / 1.95	19-1 / 2.11	19-7 / 2.27	20-0 / 2.43		19.2	
14-4 / 1.10	14-10 / 1.22	15-3 / 1.35	15-9 / 1.48	16-3 / 1.61	16-8 / 1.75	17-1 / 1.89	17-6 / 2.03	17-11 / 2.18	18-9 / 2.48	24.0	
25-10 / 1.56	26-8 / 1.73	27-7 / 1.91	28-5 / 2.09	29-3 / 2.28	30-1 / 2.47					12.0	
24-2 / 1.46	25-0 / 1.62	25-10 / 1.78	26-7 / 1.95	27-4 / 2.13	28-1 / 2.31	28-10 / 2.49				13.7	
22-4 / 1.35	23-2 / 1.50	23-11 / 1.65	24-7 / 1.81	25-4 / 1.97	26-0 / 2.14	26-8 / 2.31	27-4 / 2.48			16.0	2x10
20-5 / 1.23	21-1 / 1.37	21-10 / 1.51	22-6 / 1.65	23-2 / 1.80	23-9 / 1.95	24-5 / 2.11	25-0 / 2.27	25-7 / 2.43		19.2	
18-3 / 1.10	18-11 / 1.22	19-6 / 1.35	20-1 / 1.48	20-8 / 1.61	21-3 / 1.75	21-10 / 1.89	22-4 / 2.03	22-10 / 2.18	23-11 / 2.48	24.0	
31-4 / 1.56	32-6 / 1.73	33-6 / 1.91	34-7 / 2.09	35-7 / 2.28	36-7 / 2.47					12.0	
29-4 / 1.46	30-5 / 1.62	31-4 / 1.78	32-4 / 1.95	33-3 / 2.13	34-2 / 2.31	35-1 / 2.49				13.7	
27-2 / 1.35	28-2 / 1.50	29-1 / 1.65	29-11 / 1.81	30-10 / 1.97	31-8 / 2.14	32-6 / 2.31	33-3 / 2.48			16.0	2x12
24-10 / 1.23	25-8 / 1.37	26-6 / 1.51	27-4 / 1.65	28-2 / 1.80	28-11 / 1.95	29-8 / 2.11	30-5 / 2.27	31-1 / 2.43		19.2	
22-2 / 1.10	23-0 / 1.22	23-9 / 1.35	24-5 / 1.48	25-2 / 1.61	25-10 / 1.75	26-6 / 1.89	27-2 / 2.03	27-10 / 2.18	29-1 / 2.48	24.0	

Note: The required modulus of elasticity, "E", in 1,000,000 pounds per inch is shown below each span.

HIGH SLOPE RAFTERS

Slope over 3 in 12 - 20 Lbs. Per Sq. Ft. Live Load
(Heavy Roof Covering)

DESIGN CRITERIA:
Strength - 15 lbs per sq. ft. dead load plus 20 lbs. per sq. ft. live load determines required fiber stress.
Deflection - For 20 lbs. per sq. ft. live load. Limited to span in inches divided by 180.

RAFTER SIZE (IN)	SPACING (IN)	Allowable Extreme Fiber Stress in Bending, "F_b" (psi).											
		200	300	400	500	600	700	800	900	1000	1100	1200	1300
2x4	12.0	3-5 0.05	4-2 0.09	4-10 0.14	5-5 0.20	5-11 0.26	6-5 0.33	6-10 0.40	7-3 0.48	7-8 0.56	8-0 0.65	8-4 0.74	8-8 0.83
	13.7	3-2 0.05	3-11 0.09	4-6 0.13	5-1 0.19	5-6 0.24	6-0 0.31	6-5 0.38	6-9 0.45	7-2 0.52	7-6 0.61	7-10 0.69	8-2 0.78
	16.0	2-11 0.05	3-7 0.08	4-2 0.12	4-8 0.17	5-1 0.23	5-6 0.28	5-11 0.35	6-3 0.41	6-7 0.49	6-11 0.56	7-3 0.64	7-6 0.72
	19.2	2-8 0.04	3-4 0.07	3-10 0.11	4-3 0.16	4-8 0.21	5-1 0.26	5-5 0.32	5-9 0.38	6-0 0.44	6-4 0.51	6-7 0.58	6-11 0.66
	24.0	2-5 0.04	2-11 0.07	3-5 0.10	3-10 0.14	4-2 0.18	4-6 0.23	4-10 0.28	5-1 0.34	5-5 0.40	5-8 0.46	5-11 0.52	6-2 0.59
2x6	12.0	5-4 0.05	6-7 0.09	7-7 0.14	8-6 0.20	9-4 0.26	10-0 0.33	10-9 0.40	11-5 0.48	12-0 0.56	12-7 0.65	13-2 0.74	13-8 0.83
	13.7	5-0 0.05	6-2 0.09	7-1 0.13	7-11 0.19	8-8 0.24	9-5 0.31	10-0 0.38	10-8 0.45	11-3 0.52	11-9 0.61	12-4 0.69	12-10 0.78
	16.0	4-8 0.04	5-8 0.08	6-7 0.12	7-4 0.17	8-1 0.23	8-8 0.28	9-4 0.35	9-10 0.41	10-5 0.49	10-11 0.56	11-5 0.64	11-10 0.72
	19.2	4-3 0.04	5-2 0.07	6-0 0.11	6-9 0.16	7-4 0.21	7-11 0.26	8-6 0.32	9-0 0.38	9-6 0.44	9-11 0.51	10-5 0.58	10-10 0.66
	24.0	3-10 0.04	4-8 0.07	5-4 0.10	6-0 0.14	6-7 0.18	7-1 0.23	7-7 0.28	8-1 0.34	8-6 0.40	8-11 0.46	9-4 0.52	9-8 0.59
2x8	12.0	7-1 0.05	8-8 0.09	10-0 0.14	11-2 0.20	12-3 0.26	13-3 0.33	14-2 0.40	15-0 0.48	15-10 0.56	16-7 0.65	17-4 0.74	18-0 0.83
	13.7	6-7 0.05	8-1 0.09	9-4 0.13	10-6 0.19	11-6 0.24	12-5 0.31	13-3 0.38	14-0 0.45	14-10 0.52	15-6 0.61	16-3 0.69	16-10 0.78
	16.0	6-2 0.04	7-6 0.08	8-8 0.12	9-8 0.17	10-7 0.23	11-6 0.28	12-3 0.35	13-0 0.41	13-8 0.49	14-4 0.56	15-0 0.64	15-7 0.72
	19.2	5-7 0.04	6-10 0.07	7-11 0.11	8-10 0.16	9-8 0.21	10-6 0.26	11-2 0.32	11-10 0.38	12-6 0.44	13-1 0.51	13-8 0.58	14-3 0.66
	24.0	5-0 0.04	6-2 0.07	7-1 0.10	7-11 0.14	8-8 0.18	9-4 0.23	10-0 0.28	10-7 0.34	11-2 0.40	11-9 0.46	12-3 0.52	12-9 0.59
2x10	12.0	9-0 0.05	11-1 0.09	12-9 0.14	14-3 0.20	15-8 0.26	16-11 0.33	18-1 0.40	19-2 0.48	20-2 0.56	21-2 0.65	22-1 0.74	23-0 0.83
	13.7	8-5 0.05	10-4 0.09	11-11 0.13	13-4 0.19	14-8 0.24	15-10 0.31	16-11 0.38	17-11 0.45	18-11 0.52	19-10 0.61	20-8 0.69	21-6 0.78
	16.0	7-10 0.04	9-7 0.08	11-1 0.12	12-4 0.17	13-6 0.23	14-8 0.28	15-8 0.35	16-7 0.41	17-6 0.49	18-4 0.56	19-2 0.64	19-11 0.72
	19.2	7-2 0.04	8-9 0.07	10-1 0.11	11-3 0.16	12-4 0.21	13-4 0.26	14-3 0.32	15-2 0.38	15-11 0.44	16-9 0.51	17-6 0.58	18-2 0.66
	24.0	6-5 0.04	7-10 0.07	9-0 0.10	10-1 0.14	11-1 0.18	11-11 0.23	12-9 0.28	13-6 0.34	14-3 0.40	15-0 0.46	15-8 0.52	16-3 0.59

Note: The required modulus of elasticity, "E", in 1,000,000 pounds per square inch is shown below each span.

RAFTERS: Spans are measured along the horizontal projection and loads are considered as applied on the horizontal projection.

Allowable Extreme Fiber Stress in Bending, "Fb" (psi).												RAFTER SPACING (IN)	SIZE (IN)
1400	1500	1600	1700	1800	1900	2000	2100	2200	2400	2700	3000		
9-0 0.93	9-4 1.03	9-8 1.14	9-11 1.24	10-3 1.36	10-6 1.47	10-10 1.59	11-1 1.71	11-4 1.83	11-10 2.09	12-7 2.49		12.0	
8-5 0.87	8-9 0.96	9-0 1.06	9-4 1.16	9-7 1.27	9-10 1.37	10-1 1.48	10-4 1.60	10-7 1.71	11-1 1.95	11-9 2.33		13.7	
7-10 0.80	8-1 0.89	8-4 0.98	8-7 1.08	8-10 1.17	9-1 1.27	9-4 1.37	9-7 1.48	9-10 1.59	10-3 1.81	10-10 2.16	11-5 2.53	16.0	2x4
7-2 0.73	7-5 0.81	7-8 0.90	7-10 0.98	8-1 1.07	8-4 1.16	8-6 1.25	8-9 1.35	8-11 1.45	9-4 1.65	9-11 1.97	10-5 2.31	19.2	
6-5 0.66	6-7 0.73	6-10 0.80	7-0 0.88	7-3 0.96	7-5 1.04	7-8 1.12	7-10 1.21	8-0 1.29	8-4 1.48	8-10 1.76	9-4 2.06	24.0	
14-2 0.93	14-8 1.03	15-2 1.14	15-8 1.24	16-1 1.36	16-7 1.47	17-0 1.59	17-5 1.71	17-10 1.83	18-7 2.09	19-9 2.49		12.0	
13-3 0.87	13-9 0.96	14-2 1.06	14-8 1.16	15-1 1.27	15-6 1.37	15-11 1.48	16-3 1.60	16-8 1.71	17-5 1.95	18-5 2.33		13.7	
12-4 0.80	12-9 0.89	13-2 0.98	13-7 1.08	13-11 1.17	14-4 1.27	14-8 1.37	15-1 1.48	15-5 1.59	16-1 1.81	17-1 2.16	18-0 2.53	16.0	2x6
11-3 0.73	11-7 0.81	12-0 0.90	12-4 0.98	12-9 1.07	13-1 1.16	13-5 1.25	13-9 1.35	14-1 1.45	14-8 1.65	15-7 1.97	16-5 2.31	19.2	
10-0 0.66	10-5 0.73	10-9 0.80	11-1 0.88	11-5 0.96	11-8 1.04	12-0 1.12	12-4 1.21	12-7 1.29	13-2 1.48	13-11 1.76	14-8 2.06	24.0	
18-9 0.93	19-5 1.03	20-0 1.14	20-8 1.24	21-3 1.36	21-10 1.47	22-4 1.59	22-11 1.71	23-6 1.83	24-6 2.09	26-0 2.49		12.0	
17-6 0.87	18-2 0.96	18-9 1.06	19-4 1.16	19-10 1.27	20-5 1.37	20-11 1.48	21-5 1.60	21-11 1.71	22-11 1.95	24-4 2.33		13.7	
16-3 0.80	16-9 0.89	17-4 0.98	17-10 1.08	18-5 1.17	18-11 1.27	19-5 1.37	19-10 1.48	20-4 1.59	21-3 1.81	22-6 2.16	23-9 2.53	16.0	2x8
14-10 0.73	15-4 0.81	15-10 0.90	16-4 0.98	16-9 1.07	17-3 1.16	17-8 1.25	18-2 1.35	18-7 1.45	19-5 1.65	20-7 1.97	21-8 2.31	19.2	
13-3 0.66	13-8 0.73	14-2 0.80	14-7 0.88	15-0 0.96	15-5 1.04	15-10 1.12	16-3 1.21	16-7 1.29	17-4 1.48	18-5 1.76	19-5 2.06	24.0	
23-11 0.93	24-9 1.03	25-6 1.14	26-4 1.24	27-1 1.36	27-10 1.47	28-7 1.59	29-3 1.71	29-11 1.83	31-3 2.09	33-2 2.49		12.0	
22-4 0.87	23-2 0.96	23-11 1.06	24-7 1.16	25-4 1.27	26-0 1.37	26-8 1.48	27-4 1.60	28-0 1.71	29-3 1.95	31-0 2.33		13.7	
20-8 0.80	21-5 0.89	22-1 0.98	22-10 1.08	23-5 1.17	24-1 1.27	24-9 1.37	25-4 1.48	25-11 1.59	27-1 1.81	28-9 2.16	30-3 2.53	16.0	2x10
18-11 0.73	19-7 0.81	20-2 0.90	20-10 0.98	21-5 1.07	22-0 1.16	22-7 1.25	23-2 1.35	23-8 1.45	24-9 1.65	26-3 1.97	27-8 2.31	19.2	
16-11 0.66	17-6 0.73	18-1 0.80	18-7 0.88	19-2 0.96	19-8 1.04	20-2 1.12	20-8 1.21	22-1 1.29	23-5 1.48	24-9 1.76		24.0	

Note: The required modulus of elasticity, "E", in 1,000,000 pounds per square inch is shown below each span.

HIGH SLOPE RAFTERS
Slope over 3 in 12 · 30 Lbs. Per Sq. Ft. Live Load
(Heavy Roof Covering)

DESIGN CRITERIA:
Strength · 15 lbs. per sq. ft. dead load plus
30 lbs. per sq. ft. live load determines
required fiber stress.
Deflection · For 30 lbs. per sq. ft. live load.
Limited to span in inches divided by 180.

RAFTER SIZE (IN)	SPACING (IN)	Allowable Extreme Fiber Stress in Bending, "F_b" (psi).											
		200	300	400	500	600	700	800	900	1000	1100	1200	1300
2x4	12.0	3-0 / 0.05	3-8 / 0.09	4-3 / 0.15	4-9 / 0.20	5-3 / 0.27	5-8 / 0.34	6-0 / 0.41	6-5 / 0.49	6-9 / 0.58	7-1 / 0.67	7-5 / 0.76	7-8 / 0.86
	13.7	2-10 / 0.05	3-5 / 0.09	4-0 / 0.14	4-5 / 0.19	4-11 / 0.25	5-3 / 0.32	5-8 / 0.39	6-0 / 0.46	6-4 / 0.54	6-7 / 0.62	6-11 / 0.71	7-2 / 0.80
	16.0	2-7 / 0.04	3-2 / 0.08	3-8 / 0.13	4-1 / 0.18	4-6 / 0.23	4-11 / 0.29	5-3 / 0.36	5-6 / 0.43	5-10 / 0.50	6-1 / 0.58	6-5 / 0.66	6-8 / 0.74
	19.2	2-5 / 0.04	2-11 / 0.08	3-4 / 0.12	3-9 / 0.16	4-1 / 0.21	4-5 / 0.27	4-9 / 0.33	5-1 / 0.39	5-4 / 0.46	5-7 / 0.53	5-10 / 0.60	6-1 / 0.68
	24.0	2-2 / 0.04	2-7 / 0.07	3-0 / 0.10	3-4 / 0.14	3-8 / 0.19	4-0 / 0.24	4-3 / 0.29	4-6 / 0.35	4-9 / 0.41	5-0 / 0.47	5-3 / 0.54	5-5 / 0.61
2x6	12.0	4-9 / 0.05	5-10 / 0.09	6-8 / 0.15	7-6 / 0.20	8-2 / 0.27	8-10 / 0.34	9-6 / 0.41	10-0 / 0.49	10-7 / 0.58	11-1 / 0.67	11-7 / 0.76	12-1 / 0.86
	13.7	4-5 / 0.05	5-5 / 0.09	6-3 / 0.14	7-0 / 0.19	7-8 / 0.25	8-3 / 0.32	8-10 / 0.39	9-5 / 0.46	9-11 / 0.54	10-5 / 0.62	10-10 / 0.71	11-3 / 0.80
	16.0	4-1 / 0.04	5-0 / 0.08	5-10 / 0.13	6-6 / 0.18	7-1 / 0.23	7-8 / 0.29	8-2 / 0.36	8-8 / 0.43	9-2 / 0.50	9-7 / 0.58	10-0 / 0.66	10-5 / 0.74
	19.2	3-9 / 0.04	4-7 / 0.08	5-4 / 0.12	5-11 / 0.16	6-6 / 0.21	7-0 / 0.27	7-6 / 0.33	7-11 / 0.39	8-4 / 0.46	8-9 / 0.53	9-2 / 0.60	9-6 / 0.68
	24.0	3-4 / 0.04	4-1 / 0.07	4-9 / 0.10	5-4 / 0.14	5-10 / 0.19	6-3 / 0.24	6-8 / 0.29	7-1 / 0.35	7-6 / 0.41	7-10 / 0.47	8-2 / 0.54	8-6 / 0.61
2x8	12.0	6-3 / 0.05	7-8 / 0.09	8-10 / 0.15	9-10 / 0.20	10-10 / 0.27	11-8 / 0.34	12-6 / 0.41	13-3 / 0.49	13-11 / 0.58	14-8 / 0.67	15-3 / 0.76	15-11 / 0.86
	13.7	5-10 / 0.05	7-2 / 0.09	8-3 / 0.14	9-3 / 0.19	10-1 / 0.25	10-11 / 0.32	11-8 / 0.39	12-5 / 0.46	13-1 / 0.54	13-8 / 0.62	14-4 / 0.71	14-11 / 0.80
	16.0	5-5 / 0.04	6-7 / 0.08	7-8 / 0.13	8-7 / 0.18	9-4 / 0.23	10-1 / 0.29	10-10 / 0.36	11-6 / 0.43	12-1 / 0.50	12-8 / 0.58	13-3 / 0.66	13-9 / 0.74
	19.2	4-11 / 0.04	6-1 / 0.08	7-0 / 0.12	7-10 / 0.16	8-7 / 0.21	9-3 / 0.27	9-10 / 0.33	10-6 / 0.39	11-0 / 0.46	11-7 / 0.53	12-1 / 0.60	12-7 / 0.68
	24.0	4-5 / 0.04	5-5 / 0.07	6-3 / 0.10	7-0 / 0.14	7-8 / 0.19	8-3 / 0.24	8-10 / 0.29	9-4 / 0.35	9-10 / 0.41	10-4 / 0.47	10-10 / 0.54	11-3 / 0.61
2x10	12.0	8-0 / 0.05	9-9 / 0.09	11-3 / 0.15	12-7 / 0.20	13-9 / 0.27	14-11 / 0.34	15-11 / 0.41	16-11 / 0.49	17-10 / 0.58	18-8 / 0.67	19-6 / 0.76	20-4 / 0.86
	13.7	7-5 / 0.05	9-1 / 0.09	10-6 / 0.14	11-9 / 0.19	12-11 / 0.25	13-11 / 0.32	14-11 / 0.39	15-10 / 0.46	16-8 / 0.54	17-6 / 0.62	18-3 / 0.71	19-0 / 0.80
	16.0	6-11 / 0.04	8-5 / 0.08	9-9 / 0.13	10-11 / 0.18	11-11 / 0.23	12-11 / 0.29	13-9 / 0.36	14-8 / 0.43	15-5 / 0.50	16-2 / 0.58	16-11 / 0.66	17-7 / 0.74
	19.2	6-4 / 0.04	7-8 / 0.08	8-11 / 0.12	9-11 / 0.16	10-11 / 0.21	11-9 / 0.27	12-7 / 0.33	13-4 / 0.39	14-1 / 0.46	14-9 / 0.53	15-5 / 0.60	16-1 / 0.68
	24.0	5-8 / 0.04	6-11 / 0.07	8-0 / 0.10	8-11 / 0.14	9-9 / 0.19	10-6 / 0.24	11-3 / 0.29	11-11 / 0.35	12-7 / 0.41	13-2 / 0.47	13-9 / 0.54	14-4 / 0.61

Note: The required modulus of elasticity, "E", in 1,000,000 pounds per inch is shown below each span.

RAFTERS: Spans are measured along the horizontal projection and loads are considered as applied on the horizontal projection.

| Allowable Extreme Fiber Stress in Bending, "Fb" (psi). | | | | | | | | | | | | RAFTER | |
1400	1500	1600	1700	1800	1900	2000	2100	2200	2400	2700	3000	SPACING (IN)	SIZE (IN)
8-0 0.96	8-3 1.06	8-6 1.17	8-9 1.28	9-0 1.39	9-3 1.51	9-6 1.63	9-9 1.76	10-0 1.88	10-5 2.15	11-1 2.56		12.0	
7-5 0.89	7-9 0.99	8-0 1.09	8-3 1.20	8-5 1.30	8-8 1.41	8-11 1.53	9-2 1.64	9-4 1.76	9-9 2.01	10-4 2.40		13.7	
6-11 0.83	7-2 0.92	7-5 1.01	7-7 1.11	7-10 1.21	8-0 1.31	8-3 1.41	8-5 1.52	8-8 1.63	9-0 1.86	9-7 2.22	10-1 2.60	16.0	2x4
6-4 0.76	6-6 0.84	6-9 0.92	6-11 1.01	7-2 1.10	7-4 1.20	7-6 1.29	7-9 1.39	7-11 1.49	8-3 1.70	8-9 2.03	9-3 2.37	19.2	
5-8 0.68	5-10 0.75	6-0 0.83	6-3 0.90	6-5 0.99	6-7 1.07	6-9 1.15	6-11 1.24	7-1 1.33	7-5 1.52	7-10 1.81	8-3 2.12	24.0	
12-6 0.96	13-0 1.06	13-5 1.17	13-10 1.28	14-2 1.39	14-7 1.51	15-0 1.63	15-4 1.76	15-8 1.88	16-5 2.15	17-5 2.56		12.0	
11-9 0.89	12-2 0.99	12-6 1.09	12-11 1.20	13-3 1.30	13-8 1.41	14-0 1.53	14-4 1.64	14-8 1.76	15-4 2.01	16-3 2.40		13.7	
10-10 0.83	11-3 0.92	11-7 1.01	11-11 1.11	12-4 1.21	12-8 1.31	13-0 1.41	13-3 1.52	13-7 1.63	14-2 1.86	15-1 2.22	15-11 2.60	16.0	2x6
9-11 0.76	10-3 0.84	10-7 0.92	10-11 1.01	11-3 1.10	11-6 1.20	11-10 1.29	12-2 1.39	12-5 1.49	13-0 1.70	13-9 2.03	14-6 2.37	19.2	
8-10 0.68	9-2 0.75	9-6 0.83	9-9 0.90	10-0 0.99	10-4 1.07	10-7 1.15	10-10 1.24	11-1 1.33	11-7 1.52	12-4 1.81	13-0 2.12	24.0	
16-6 0.96	17-1 1.06	17-8 1.17	18-2 1.28	18-9 1.39	19-3 1.51	19-9 1.63	20-3 1.76	20-8 1.88	21-7 2.15	22-11 2.56		12.0	
15-5 0.89	16-0 0.99	16-6 1.09	17-0 1.20	17-6 1.30	18-0 1.41	18-5 1.53	18-11 1.64	19-4 1.76	20-3 2.01	21-5 2.40		13.7	
14-4 0.83	14-10 0.92	15-3 1.01	15-9 1.11	16-3 1.21	16-8 1.31	17-1 1.41	17-6 1.52	17-11 1.63	18-9 1.86	19-10 2.22	20-11 2.60	16.0	2x8
13-1 0.76	13-6 0.84	13-11 0.92	14-5 1.01	14-10 1.10	15-2 1.20	15-7 1.29	16-0 1.39	16-4 1.49	17-1 1.70	18-2 2.03	19-1 2.37	19.2	
11-8 0.68	12-1 0.75	12-6 0.83	12-10 0.90	13-3 0.99	13-7 1.07	13-11 1.15	14-4 1.24	14-8 1.33	15-3 1.52	16-3 1.81	17-1 2.12	24.0	
21-1 0.96	21-10 1.06	22-6 1.17	23-3 1.28	23-11 1.39	24-6 1.51	25-2 1.63	25-10 1.76	26-5 1.88	27-7 2.15	29-3 2.56		12.0	
19-8 0.89	20-5 0.99	21-1 1.09	21-9 1.20	22-4 1.30	22-11 1.41	23-7 1.53	24-2 1.64	24-8 1.76	25-10 2.01	27-4 2.40		13.7	
18-3 0.83	18-11 0.92	19-6 1.01	20-1 1.11	20-8 1.21	21-3 1.31	21-10 1.41	22-4 1.52	22-10 1.63	23-11 1.86	25-4 2.22	26-8 2.60	16.0	2x10
16-8 0.76	17-3 0.84	17-10 0.92	18-4 1.01	18-11 1.10	19-5 1.20	19-11 1.29	20-5 1.39	20-10 1.49	21-10 1.70	23-2 2.03	24-5 2.37	19.2	
14-11 0.68	15-5 0.75	15-11 0.83	16-5 0.90	16-11 0.99	17-4 1.07	17-10 1.15	18-3 1.24	18-8 1.33	19-6 1.52	20-8 1.81	21-10 2.12	24.0	

Note: The required modulus of elasticity, "E", in 1,000,000 pounds per square inch is shown below each span.

Maximum Spans
for Joists and Rafters

(Courtesy of Southern Forest Products Association)

TABLE NO. 1. Floor Joists—30 psf live load. Sleeping rooms and attic floors. (Spans shown in lightface type are based on a deflection limitation of $l/360$. Spans shown in **boldface type** are limited by allowable extreme fiber stress in bending of the grade and include a 10 psf dead load.)

GRADE

SIZE*	O.C.†	Dense Sel Str KD and No. 1 Dense KD	Dense Sel Str and Sel Str KD	Appearance KD, No. 1 Dense and No. 1 KD	Sel Str, No. 1 and No. 2 Dense KD	No. 2 Dense and No. 2 MG KD	No. 2 MG
2 × 6	12.0	12-9	12-6	12-6	12-3	12-0	11-10
	16.0	11-7	11-4	11-4	11-2	10-11	10-9
	24.0	10-1	9-11	9-1	9-9	9-7	9-4
2 × 8	12.0	16-9	16-6	16-6	16-2	15-10	15-7
	16.0	15-3	15-0	15-0	14-8	14-5	14-2
	24.0	13-4	13-1	13-1	12-10	12-7	12-4
2 × 10	12.0	21-5	21-0	21-0	20-8	20-3	19-10
	16.0	19-5	19-1	19-1	18-9	18-5	18-0
	24.0	17-0	16-8	16-8	16-5	16-1	15-9
2 × 12	12.0	26-0	25-7	25-7	25-1	24-8	24-2
	16.0	23-7	23-3	23-3	22-10	22-5	21-11
	24.0	20-8	20-3	20-3	19-11	19-7	19-2

SIZE*	O.C.†	No. 2 KD	No. 2	No. 3 Dense KD	No. 3 Dense	No. 3 KD	No. 3
2 × 6	12.0	11-7	11-3	**11-6**	**11-0**	10-9	**10-2**
	16.0	10-6	10-3	**9-11**	**9-7**	9-4	**8-9**
	24.0	**9-0**	8-8	8-1	**7-10**	7-7	7-2
2 × 8	12.0	15-3	14-11	**15-2**	**14-7**	14-4	**13-5**
	16.0	13-10	13-6	**13-1**	**12-8**	12-3	**11-7**
	24.0	**11-11**	**11-6**	10-9	**10-3**	10-0	9-6
2 × 10	12.0	19-5	19-0	**19-4**	**18-7**	18-1	**17-2**
	16.0	17-8	17-3	**16-9**	**16-2**	15-7	**14-10**
	24.0	**15-3**	**14-7**	13-8	**13-2**	12-9	12-2
2 × 12	12.0	23-7	23-1	**23-5**	**22-8**	22-1	**20-10**
	16.0	21-6	21-0	**19-4**	**19-7**	19-1	**18-1**
	24.0	**18-6**	**17-9**	16-8	**16-0**	15-7	**14-9**

*Size = nominal lumber size given in inches
†O.C. = on-center spacing given in inches

TABLE NO. 2. Floor Joists—40 psf live load. All rooms except sleeping rooms and attic floors. (Spans shown in lightface type are based on a deflection limitation of $l/360$. Spans shown in **boldface type** are limited by allowable extreme fiber stress in bending of the grade and include a 10 psf dead load.)

GRADE

SIZE*	O.C.†	Dense Sel Str KD and No. 1 Dense KD	Dense Sel Str and Sel Str KD	Appearance KD, No. 1 Dense and No. 1 KD	Sel Str, No. 1 and No. 2 Dense KD	No. 2 Dense and No. 2 MG KD	No. 2 MG
2 × 6	12.0	11-7	11-4	11-4	11-2	10-11	10-9
	16.0	10-6	10-4	10-4	10-2	9-11	9-9
	24.0	9-2	9-0	9-0	8-10	8-8	8-6
2 × 8	12.0	15-3	15-0	15-0	14-8	14-5	14-2
	16.0	13-10	13-7	13-7	13-4	13-1	12-10
	24.0	12-1	11-11	11-11	11-8	11-5	11-3
2 × 10	12.0	19-5	19-1	19-1	18-9	18-5	18-0
	16.0	17-8	17-4	17-4	17-0	16-9	16-5
	24.0	15-5	15-2	15-2	14-11	14-7	14-4
2 × 12	12.0	23-7	23-3	23-3	22-10	22-5	21-11
	16.0	21-6	21-1	21-1	20-9	20-4	19-11
	24.0	18-9	18-5	18-5	18-1	17-9	17-5

SIZE*	O.C.†	No. 2 KD	No. 2	No. 3 Dense KD	No. 3 Dense	No. 3 KD	No. 3
2 × 6	12.0	10-6	10-3	**10-3**	9-11	9-8	9-1
	16.0	9-6	9-4	**8-10**	8-6	8-4	7-10
	24.0	**8-1**	7-9	7-3	6-11	6-9	6-6
2 × 8	12.0	13-10	13-6	**13-6**	13-0	12-8	12-0
	16.0	12-7	12-3	**11-8**	11-3	11-0	10-5
	24.0	**10-8**	10-3	9-7	9-2	9-0	8-7
2 × 10	12.0	17-8	17-3	**17-3**	16-8	16-3	15-4
	16.0	16-0	15-8	**14-11**	14-5	14-0	13-2
	24.0	**13-7**	13-1	12-3	11-9	11-5	10-10
2 × 12	12.0	21-6	21-0	**21-0**	20-3	19-8	18-8
	16.0	19-6	19-1	**18-1**	17-6	17-0	16-1
	24.0	**16-6**	15-11	14-11	14-3	13-11	13-4

*Size = nominal lumber size given in inches
†O.C. = on-center spacing given in inches

TABLE NO. 3. Ceiling Joists—Drywall Ceiling—20 psf live load. No future sleeping rooms but limited storage available. (Spans shown in lightface type are based on a deflection limitation of $l/240$. Spans shown in **boldface type** are limited by allowable extreme fiber stress in bending of the grade and include a 10 psf dead load.)

GRADE

SIZE*	O.C.†	Dense Sel Str KD and No. 1 Dense KD	Dense Sel Str and Sel Str KD	Appearance KD, No. 1 Dense and No. 1 KD	Sel Str, No. 1 and No. 2 Dense KD	No. 2 Dense and No. 2 MG KD	No. 2 MG	No. 2 KD	No. 2
2 × 4	12.0	10-7	10-5	10-5	10-3	10-0	9-10	9-8	9-5
	16.0	9-8	9-6	9-6	9-4	9-1	8-11	8-9	8-7
	24.0	8-5	8-3	8-3	8-1	**7-10**	**7-6**	**7-1**	**7-0**
2 × 6	12.0	16-8	16-4	16-4	16-1	15-9		14-9	14-2
	16.0	15-2	14-11	14-11	14-7	**13-11**	**13-6**	12-9	12-3
	24.0	13-3	13-0	13-0	12-9	**12-0**	**11-0**	10-5	10-0
2 × 8	12.0	21-11	21-7	21-7	21-2	20-10		19-6	18-8
	16.0	19-11	19-7	19-7	19-3	**18-4**	**17-10**	16-10	16-2
	24.0	17-5	17-2	17-2	16-10	**15-10**	**14-6**	13-9	13-2
2 × 10	12.0	28-0	27-6	27-6	27-1	26-6		24-10	23-11
	16.0	25-5	25-0	25-0	24-7	**23-6**	**22-9**	21-7	20-7
	24.0	22-3	21-10	21-10	21-6	**19-2**	**18-6**	17-7	**16-10**

SIZE*	O.C.†	No. 3 Dense KD	No. 3 Dense	No. 3 KD	No. 3	Construction KD	Construction	Standard KD	Standard
2 × 4	12.0	**9-0**	**8-7**	**8-2**	**8-0**	**9-5**	**9-0**	**7-1**	**6-11**
	16.0	**7-9**	**7-6**	**7-1**	**6-11**	**8-1**	**7-9**	**6-2**	**5-11**
	24.0	**6-4**	**6-1**	**5-9**	**5-8**	**6-7**	**6-4**	**5-1**	**4-10**
2 × 6	12.0	**13-3**	**12-9**	**12-5**	**11-9**				
	16.0	**11-5**	**11-0**	**10-9**	**10-4**				
	24.0	**9-4**	**9-0**	**8-9**	**8-4**				
2 × 8	12.0	**17-5**	**16-9**	**16-4**	**15-6**				
	16.0	**15-0**	**14-6**	**14-2**	**13-5**				
	24.0	**12-4**	**11-11**	**11-8**	**11-0**				
2 × 10	12.0	**22-3**	**21-6**	**20-11**	**19-9**				
	16.0	**19-3**	**18-6**	**18-1**	**17-2**				
	24.0	**15-10**	**15-4**	**14-11**	**14-0**				

*Size = nominal lumber size given in inches
†O.C. = on-center spacing given in inches

TABLE NO. 4. Ceiling Joists—Drywall Ceiling—10 psf live load. No future sleeping rooms and no attic storage, roof slopes 3 in 12 or less. (Spans shown in lightface type are based on a deflection limitation of $l/240$. Spans shown in **boldface type** are limited by allowable extreme fiber stress in bending of the grade and include a 5 psf dead load.).

GRADE

SIZE*	O.C.†	Dense Sel Str KD and No. 1 Dense KD	Dense Sel Str and Sel Str KD	Appearance KD, No. 1 Dense and No. 1 KD	Sel Str, No. 1 and No. 2 Dense KD	No. 2 Dense and No. 2 MG KD	No. 2 MG	No. 2 KD	No. 2
2 × 4	12.0	13-4	13-2	13-2	12-11	12-8	12-5	12-2	11-10
	16.0	12-2	11-11	11-11	11-9	11-6	11-3	11-0	10-9
	24.0	10-7	10-5	10-5	10-3	10-0	9-10	9-8	9-5
2 × 6	12.0	21-0	20-8	20-8	20-3	19-11	19-6	19-1	18-8
	16.0	19-1	18-9	18-9	18-5	18-1	17-8	17-4	16-11
	24.0	16-8	16-4	16-4	16-1	15-9	15-6	**14-9**	**14-2**
2 × 8	12.0	27-8	27-2	27-2	26-9	26-2	25-8	25-2	24-7
	16.0	25-2	24-8	24-8	24-3	23-10	23-4	22-10	22-4
	24.0	21-11	21-7	21-7	21-2	20-10	20-5	**19-6**	**18-8**
2 × 10	12.0	35-4	34-8	34-8	34-1	33-5	32-9	32-1	31-4
	16.0	32-1	31-6	31-6	31-0	30-5	29-9	29-2	28-6
	24.0	28-0	27-6	27-6	27-1	26-6	26-0	**24-10**	**23-11**

SIZE*	O.C.†	No. 3 Dense KD	No. 3 Dense	No. 3 KD	No. 3	Construction KD	Construction	Standard KD	Standard
2 × 4	12.0	12-5	12-2	**11-8**	**11-4**	12-2	11-10	**10-1**	9-9
	16.0	**10-11**	**10-8**	10-2	9-10	11-0	10-9	**8-8**	**8-5**
	24.0	**8-11**	8-7	8-3	8-0	**9-5**	**9-0**	7-2	6-11
2 × 6	12.0	**18-7**	**18-1**	17-7	**16-7**				
	16.0	**16-3**	**15-7**	15-3	**14-4**				
	24.0	**13-3**	12-9	12-5	11-9				
2 × 8	12.0	**24-9**	23-11	23-3	21-10				
	16.0	**21-4**	20-7	20-1	18-10				
	24.0	**17-6**	16-10	16-4	15-6				
2 × 10	12.0	**31-7**	30-6	29-7	28-0				
	16.0	**27-4**	26-3	25-7	24-2				
	24.0	**22-4**	21-6	21-0	19-9				

*Size = nominal lumber size given in inches
†O.C. = on-center spacing given in inches

TABLE NO. 5. Rafters—Any Slope—With Drywall Ceiling—20 psf live load. (Spans shown in lightface type are based on a deflection limitation of $l/240$. Spans shown in **boldface type** are limited by allowable extreme fiber stress in bending of the grade and include a 15 psf dead load.)

GRADE

SIZE*	O.C.†	Dense Sel Str KD	Dense Sel Str	Sel Str KD	Sel Str	No. 1 Dense KD	No. 1 Dense	Appearance KD and No. 1 KD	No. 2 Dense KD	No. 1
2 × 6	12.0	16-8	16-4	16-4	16-0	16-8	16-4	16-4	16-0	**15-10**
	16.0	15-1	14-10	14-10	14-7	15-1	**14-10**	**14-4**	**13-11**	**13-9**
	24.0	13-2	12-11	**12-9**	**12-2**	**12-7**	**12-2**	**11-8**	**11-5**	**11-3**
2 × 8	12.0	21-11	21-7	21-5	21-3	21-11	21-7	21-5	21-2	**20-11**
	16.0	19-11	19-7	19-7	19-3	19-11	**19-7**	**18-11**	**18-5**	**18-1**
	24.0	17-5	17-1	**16-9**	**16-0**	**16-7**	**16-0**	**15-5**	**15-0**	**14-9**
2 × 10	12.0	28-0	27-6	27-6	27-0	28-0	27-6	27-6	27-0	**26-8**
	16.0	25-5	25-0	25-0	24-8	25-5	**25-0**	**24-1**	**23-5**	**23-1**
	24.0	22-2	21-10	**21-7**	**20-5**	**21-2**	**20-5**	**19-8**	**19-2**	**18-10**
2 × 12	12.0	34-0	33-5	33-5	32-10	34-0	33-5	33-5	32-10	**32-5**
	16.0	30-11	30-5	30-5	29-11	30-11	**30-5**	**29-4**	**28-6**	**28-1**
	24.0	27-0	26-7	**26-0**	**24-10**	**25-9**	**24-10**	**23-11**	**23-3**	**22-11**

SIZE*	O.C.†	No. 2 Dense	No. 2 MG KD	No. 2 MG	No. 2 KD	No. 2	No. 3 Dense KD	No. 3 Dense	No. 3 KD	No. 3
2 × 6	12.0	**15-3**	**14-11**	**14-5**	**13-8**	**13-2**	**12-3**	**11-10**	**11-6**	**10-11**
	16.0	**13-4**	**12-11**	**12-6**	**11-10**	**11-5**	**10-8**	**10-3**	**9-11**	**9-6**
	24.0	**10-11**	**10-7**	**10-2**	**9-8**	**9-4**	**8-8**	**8-4**	**8-2**	**7-8**
2 × 8	12.0	**20-2**	**19-8**	**19-1**	**18-0**	**17-4**	**16-2**	**15-7**	**15-2**	**14-4**
	16.0	**17-6**	**17-0**	**16-6**	**15-7**	**15-0**	**14-0**	**13-6**	**13-2**	**12-5**
	24.0	**14-4**	**13-11**	**13-5**	**12-9**	**12-3**	**11-5**	**11-0**	**10-8**	**10-1**
2 × 10	12.0	**25-10**	**25-1**	**24-4**	**23-0**	**22-1**	**20-8**	**19-11**	**19-5**	**18-4**
	16.0	**22-4**	**21-9**	**21-0**	**19-11**	**19-2**	**17-11**	**17-3**	**16-9**	**15-10**
	24.0	**18-3**	**17-9**	**17-2**	**16-3**	**15-8**	**14-7**	**14-0**	**13-8**	**12-11**
2 × 12	12.0	**31-6**	**30-7**	**29-7**	**28-0**	**26-11**	**25-2**	**24-3**	**23-7**	**22-3**
	16.0	**27-4**	**26-5**	**25-7**	**24-3**	**23-3**	**21-9**	**20-11**	**20-5**	**19-3**
	24.0	**22-3**	**21-7**	**20-10**	**19-10**	**19-0**	**17-9**	**17-1**	**16-8**	**15-9**

*Size = nominal lumber size given in inches
†O.C. = on-center spacing given in inches

TABLE NO. 6. Rafters—Any Slope—With Drywall Ceiling—30 psf live load. (Spans shown in lightface type are based on a deflection limitation of $l/240$. Spans shown in **boldface type** are limited by allowable extreme fiber stress in bending of the grade and include a 15 psf dead load.)

GRADE

SIZE*	O.C.†	Dense Sel Str KD	Dense Sel Str	Sel Str KD	Sel Str	No. 1 Dense KD	No. 1 Dense	Appearance KD and No. 1 KD	No. 2 Dense KD	No. 1
2 × 6	12.0	14-6	14-3	14-3	14-0	14-7	14-2	14-2	14-0	**14-0**
	16.0	13-2	13-0	13-0	12-9	13-3	13-0	**12-8**	**12-4**	**12-1**
	24.0	11-6	11-4	**11-2**	**10-8**	**11-1**	**10-9**	**10-4**	**10-0**	**9-10**
2 × 8	12.0	19-2	18-10	18-10	18-6	19-3	18-9	18-9	18-6	**18-5**
	16.0	17-4	17-1	17-1	16-9	17-0	17-1	**16-8**	**16-3**	**16-0**
	24.0	15-2	14-11	**14-10**	**14-1**	**14-8**	**14-1**	**13-7**	**13-3**	**13-0**
2 × 10	12.0	24-5	24-0	24-0	23-7	24-6	23-11	23-11	23-7	**23-7**
	16.0	22-2	21-10	21-10	21-5	22-3	21-10	**21-3**	**20-8**	**20-4**
	24.0	19-5	19-1	**18-10**	**18-0**	**18-8**	**18-0**	**17-4**	**16-11**	**16-8**
2 × 12	12.0	29-9	29-2	29-2	28-9	29-9	29-1	29-1	28-9	**28-8**
	16.0	27-0	26-6	26-6	26-0	27-1	26-6	**25-10**	**25-2**	**24-9**
	24.0	23-8	23-3	**22-11**	**21-11**	**22-8**	**21-11**	**21-1**	**20-6**	**20-3**

SIZE*	O.C.†	No. 2 Dense	No. 2 MG KD	No. 2 MG	No. 2 KD	No. 2	No. 3 Dense KD	No. 3 Dense	No. 3 KD	No. 3
2 × 6	12.0	**13-7**	**13-2**	**12-8**	**12-1**	**11-7**	**10-10**	**10-5**	**10-1**	**9-7**
	16.0	**11-8**	**11-5**	**11-0**	**10-5**	**10-0**	**9-5**	**9-0**	**8-9**	**8-3**
	24.0	**9-7**	**9-3**	**9-0**	**8-6**	**8-2**	**7-8**	**7-4**	**7-2**	**6-9**
2 × 8	12.0	**17-10**	**17-4**	**16-9**	**15-11**	**15-3**	**14-3**	**13-9**	**13-5**	**12-8**
	16.0	**15-6**	**15-0**	**14-6**	**13-9**	**13-3**	**12-4**	**11-11**	**11-7**	**11-0**
	24.0	**12-8**	**12-3**	**11-9**	**11-3**	**10-10**	**10-2**	**9-8**	**9-5**	**9-0**
2 × 10	12.0	**22-7**	**22-2**	**21-4**	**20-4**	**19-6**	**18-2**	**17-7**	**17-1**	**16-2**
	16.0	**19-9**	**19-2**	**18-6**	**17-7**	**16-11**	**15-9**	**15-2**	**14-10**	**13-11**
	24.0	**16-2**	**15-8**	**15-2**	**14-4**	**13-9**	**12-10**	**12-5**	**12-1**	**11-5**
2 × 12	12.0	**27-9**	**26-11**	**26-0**	**24-8**	**23-9**	**22-2**	**21-4**	**20-9**	**19-7**
	16.0	**24-0**	**23-4**	**22-7**	**21-5**	**20-6**	**19-2**	**18-6**	**18-0**	**17-0**
	24.0	**19-8**	**19-1**	**18-5**	**17-5**	**16-9**	**15-8**	**15-1**	**14-8**	**13-10**

*Size = nominal lumber size given in inches
†O.C. = on-center spacing given in inches

TABLE NO. 7. Rafters—Any Slope—With Drywall Ceiling—40 psf live load. (Spans shown in lightface type are based on a deflection limitation of $l/240$. Spans shown in boldface type are limited by allowable extreme fiber stress in bending of the grade and include a 15 psf dead load.)

GRADE

SIZE*	O.C.†	Dense Sel Str KD	Dense Sel Str	Sel Str KD	Sel Str	No. 1 Dense KD	No. 1 Dense	Appearance KD and No. 1 KD	No. 2 Dense KD	No. 1
2 × 6	12.0	13-2	13-0	13-0	12-9	13-2	12-11	12-11	12-9	**12-8**
	16.0	12-0	11-10	11-10	11-7	12-0	11-9	**11-5**	**11-1**	**10-11**
	24.0	10-6	10-3	**10-3**	**9-8**	**10-0**	**9-8**	**9-4**	**9-1**	**8-11**
2 × 8	12.0	17-5	17-2	17-2	16-9	17-5	17-1	17-1	16-10	**16-8**
	16.0	15-10	15-6	15-6	15-3	15-10	15-6	**15-1**	**14-8**	**14-5**
	24.0	13-10	13-7	**13-6**	**12-9**	**13-3**	**12-9**	**12-4**	**12-0**	**11-10**
2 × 10	12.0	22-2	21-10	21-10	21-5	22-2	21-10	21-10	21-6	**21-3**
	16.0	20-2	19-10	19-10	19-6	20-2	19-10	**19-3**	**18-9**	**18-5**
	24.0	17-8	17-4	**16-3**	**16-3**	**16-11**	**16-3**	**15-8**	**15-3**	**15-0**
2 × 12	12.0	27-0	26-7	26-7	26-1	27-0	26-7	26-7	26-2	**25-10**
	16.0	24-7	24-1	24-1	23-9	24-7	24-2	**23-5**	**22-9**	**22-5**
	24.0	21-5	21-0	**20-11**	**19-10**	**20-6**	**19-10**	**19-1**	**18-7**	**18-4**

SIZE*	O.C.†	No. 2 Dense	No. 2 MG KD	No. 2 MG	No. 2 KD	No. 2	No. 3 Dense KD	No. 3 Dense	No. 3 KD	No. 3
2 × 6	12.0	**12-3**	**11-11**	**11-6**	**10-11**	**10-6**	**9-9**	**9-5**	**9-2**	**8-8**
	16.0	**10-8**	**10-4**	**10-0**	**9-5**	**9-1**	**8-5**	**8-1**	**7-11**	**7-6**
	24.0	**8-8**	**8-5**	**8-1**	**7-9**	**7-5**	**6-11**	**6-8**	**6-6**	**6-2**
2 × 8	12.0	**16-2**	**15-8**	**14-2**	**14-5**	**13-10**	**12-10**	**12-5**	**12-1**	**11-5**
	16.0	**14-0**	**13-7**	**12-2**	**12-6**	**12-0**	**11-2**	**10-9**	**10-5**	**9-11**
	24.0	**11-5**	**11-1**	**10-9**	**10-2**	**9-9**	**9-1**	**8-9**	**8-7**	**8-1**
2 × 10	12.0	**20-8**	**20-0**	**19-5**	**18-4**	**17-8**	**16-5**	**15-10**	**15-5**	**14-7**
	16.0	**17-11**	**17-4**	**16-9**	**15-11**	**15-3**	**14-3**	**13-9**	**13-5**	**12-8**
	24.0	**14-7**	**14-2**	**13-8**	**13-0**	**12-6**	**11-8**	**11-3**	**10-11**	**10-4**
2 × 12	12.0	**24-3**	**24-4**	**23-7**	**22-4**	**21-5**	**20-0**	**19-4**	**18-10**	**17-8**
	16.0	**21-9**	**21-1**	**20-5**	**19-4**	**18-7**	**17-4**	**16-9**	**16-3**	**15-4**
	24.0	**17-9**	**17-3**	**16-8**	**15-9**	**15-2**	**14-2**	**13-8**	**13-4**	**12-7**

*Size = nominal lumber size given in inches
†O.C. = on-center spacing given in inches

TABLE NO. 8. Rafters—Low Slope (3 in 12 or less)—With No Finished Ceiling—20 psf live load. (Spans shown in lightface type are based on a deflection limitation of $l/240$. Spans shown in **boldface type** are limited by allowable extreme fiber stress in bending of the grade and include a 10 psf dead load.)

GRADE

SIZE*	O.C.†	Dense Sel Str KD and No. 1 Dense KD	Dense Sel Str and No. 1 Dense	Sel Str KD, Appearance KD and No. 1 KD	Sel Str	No. 2 Dense KD	No. 1	No. 2 Dense	No. 2 MG KD
2 × 6	12.0	16-8	16-4	16-4	16-1	16-1	16-1	15-9	15-9
	16.0	15-2	14-11	14-11	14-8	14-8	14-8	**14-4**	**13-11**
	24.0	13-3	13-0	**12-8**	12-9	**12-4**	**12-1**	**11-9**	**11-5**
2 × 8	12.0	21-11	21-7	21-7	21-2	21-2	21-2	20-10	20-10
	16.0	19-11	19-7	19-7	19-4	19-4	19-4	**19-0**	**18-5**
	24.0	17-5	17-1	**16-8**	16-9	**16-3**	**16-0**	**15-6**	**15-0**
2 × 10	12.0	28-0	27-6	27-6	27-1	27-1	27-1	26-6	26-6
	16.0	25-5	25-0	25-0	24-7	24-7	24-7	**24-2**	**23-6**
	24.0	22-3	21-10	**21-3**	21-5	**20-8**	**20-4**	**19-9**	**19-2**
2 × 12	12.0	34-0	33-6	33-6	32-11	32-11	32-11	32-5	32-5
	16.0	30-10	30-6	30-6	29-11	29-11	29-11	**29-4**	**28-4**
	24.0	27-3	26-6	**25-10**	26-1	**25-2**	**24-9**	**24-0**	**23-4**

SIZE*	O.C.†	No. 2 MG	No. 2 KD	No. 2	No.3 Dense KD	No. 3 Dense	No. 3 KD	No. 3
2 × 6	12.0	15-6	**14-9**	**14-2**	13-3	**12-10**	**12-6**	**11-9**
	16.0	**13-6**	**12-10**	**12-4**	**11-5**	**11-1**	**10-9**	**10-4**
	24.0	**11-0**	**10-5**	**10-0**	9-4	9-0	8-9	8-4
2 × 8	12.0	20-5	**19-6**	**18-9**	17-5	16-10	**16-5**	15-6
	16.0	**17-10**	**16-10**	**16-3**	15-0	14-7	**14-2**	**13-5**
	24.0	**14-6**	13-9	**13-3**	12-4	**11-11**	**11-7**	**11-0**
2 × 10	12.0	26-0	**24-10**	**23-11**	22-3	21-6	20-11	19-9
	16.0	**22-9**	21-6	**20-8**	19-3	18-8	**18-2**	17-2
	24.0	**18-6**	17-7	**16-11**	15-10	15-2	**14-10**	**14-0**
2 × 12	12.0	**31-10**	30-3	**29-1**	27-2	26-2	25-6	**24-0**
	16.0	**27-7**	**26-2**	**25-2**	23-6	22-8	22-0	**20-9**
	24.0	**22-7**	21-5	**20-6**	19-2	18-6	18-0	**17-0**

*Size = nominal lumber size given in inches
†O.C. = on-center spacing given in inches

TABLE NO. 9. Rafters—Low Slope (3 in 12 or less)—With No Finished Ceiling—30 psf live load. (Spans shown in lightface type are based on a deflection limitation of $l/240$. Spans shown in **boldface type** are limited by allowable extreme fiber stress in bending of the grade and include a 10 psf dead load.)

GRADE

SIZE*	O.C.†	Dense Sel Str KD and No. 1 Dense KD	Dense Sel Str and No. 1 Dense	Sel Str KD, Appearance KD and No. 1 KD	Sel Str	No. 2 Dense KD	No. 1	No. 2 Dense	No. 2 MG KD
2 × 6	12.0	14-7	14-4	14-4	14-0	14-0	14-0	13-9	13-9
	16.0	13-3	13-0	13-0	12-9	12-9	12-9	**12-5**	**12-1**
	24.0	11-6	11-4	**10-11**	11-1	**10-8**	**10-6**	**10-2**	**9-10**
2 × 8	12.0	18-10	18-11	18-11	18-6	18-6	18-6	18-2	18-2
	16.0	17-4	17-1	17-1	16-10	16-10	16-10	**16-6**	**15-10**
	24.0	15 2	14-11	**14-5**	14-8	**14-0**	**13-10**	**13-5**	**13-0**
2 × 10	12.0	24-5	24-1	24-1	23-7	23-7	23-7	23-2	23-2
	16.0	21-2	21-10	21-10	21-6	21-6	21-5	**21-0**	**20-2**
	24.0	19-4	19-1	**18-5**	18-9	**17-1**	**17-8**	**17-1**	**16-7**
2 × 12	12.0	29-8	29-2	29-2	28-9	28-9	28-9	28-2	28-2
	16.0	27-0	26-7	26-7	26-1	26-1	26-1	**25-6**	**24-9**
	24.0	23-6	23-2	**22-5**	22-9	**21-9**	**21-5**	**20-10**	**20-2**

SIZE*	O.C.†	No. 2 MG	No. 2 KD	No. 2	No. 3 Dense KD	No. 3 Dense	No. 3 KD	No. 3
2 × 6	12.0	13-5	**12-10**	12-4	11-5	11-1	**10-9**	**10-2**
	16.0	**11-8**	**11-1**	10-8	9-11	9-7	**9-4**	**8-9**
	24.0	9-7	9-1	8-8	8-1	7-9	7-7	7-2
2 × 8	12.0	17-10	**16-10**	16-3	15-1	14-7	**14-2**	**13-5**
	16.0	**15-4**	14-7	14-0	13-2	12-8	**12-4**	**11-7**
	24.0	**12-7**	11-11	11-6	10-8	10-4	**10-0**	**9-6**
2 × 10	12.0	**22-8**	**21-6**	20-8	19-3	18-8	**18-2**	**17-2**
	16.0	**19-7**	**18-8**	17-11	16-8	16-1	**15-8**	**14-10**
	24.0	**16-0**	15-3	14-8	13-8	13-2	**12-10**	**12-2**
2 × 12	12.0	**27-7**	**26-2**	25-2	23-6	22-8	**22-0**	**20-10**
	16.0	**24-0**	**22-8**	21-9	20-4	19-7	**19-1**	**18-1**
	24.0	**19-6**	18-6	17-9	16-8	16-0	**15-7**	**14-9**

*Size = nominal lumber size given in inches
†O.C. = on-center spacing given in inches

TABLE NO. 10. Rafters—Low Slope (3 in 12 or less)—With No Finished Ceiling—40 psf live load. (Spans shown in lightface type are based on a deflection limitation of $l/240$. Spans shown in **boldface type** are limited by allowable extreme fiber stress in bending of the grade and include a 10 psf dead load.)

GRADE

SIZE*	O.C.†	Dense Sel Str KD and No. 1 Dense KD	Dense Sel Str and No. 1 Dense	Sel Str KD, Appearance KD and No. 1 KD	Sel Str	No. 2 Dense KD	No. 1	No. 2 Dense	No. 2 MG KD
2 × 6	12.0	13-2	12-11	12-11	12-8	12-8	12-8	12-6	**12-6**
	16.0	12-0	11-9	11-9	11-7	11-7	**11-6**	**11-2**	**10-10**
	24.0	10-5	**10-1**	9-9	10-1	**9-6**	**9-4**	**9-1**	**8-10**
2 × 8	12.0	17-5	17-1	17-1	16-10	16-10	16-10	16-6	**16-6**
	16.0	15-10	15-7	15-7	15-3	15-3	**15-2**	**14-8**	**14-3**
	24.0	13-10	**13-5**	**12-11**	13-4	**12-7**	**12-4**	**12-0**	**11-6**
2 × 10	12.0	22-1	21-10	21-10	21-5	21-5	21-5	21-0	**21-0**
	16.0	20-2	19-10	19-10	19-5	19-5	**19-4**	**18-9**	**18-2**
	24.0	17-8	**17-1**	**16-6**	17-0	**16-0**	**15-9**	**15-4**	**14-10**
2 × 12	12.0	27-0	26-6	26-6	26-1	26-1	26-1	25-7	**25-7**
	16.0	24-6	24-1	24-1	23-8	23-8	**23-6**	**22-10**	**22-1**
	24.0	21-6	**20-9**	**20-0**	20-8	**19-6**	**19-2**	**18-8**	**18-0**

SIZE*	O.C.†	No. 2 MG	No. 2 KD	No. 2	No.3 Dense KD	No. 3 Dense	No. 3 KD	No. 3
2 × 6	12.0	**12-1**	**11-5**	**11-0**	**10-3**	**9-10**	**9-7**	**9-1**
	16.0	**10-5**	**9-11**	**9-6**	**8-10**	**8-6**	**8-4**	**7-10**
	24.0	**8-6**	**8-1**	**7-9**	**7-3**	**7-0**	**6-10**	**6-6**
2 × 8	12.0	**15-11**	**15-1**	**14-6**	**13-7**	**13-1**	**12-9**	**12-0**
	16.0	**13-9**	**13-1**	**12-7**	**11-9**	**11-4**	**11-0**	**10-5**
	24.0	**11-3**	**10-8**	**10-3**	**9-7**	**9-2**	**9-0**	**8-7**
2 × 10	12.0	**20-4**	**19-3**	**18-6**	**17-3**	**16-8**	**16-2**	**15-4**
	16.0	**17-7**	**16-8**	**16-0**	**15-0**	**14-5**	**14-1**	**13-2**
	24.0	**14-5**	**13-7**	**13-1**	**12-2**	**11-9**	**11-5**	**10-10**
2 × 12	12.0	**24-9**	**23-5**	**22-6**	**21-0**	**20-3**	**19-9**	**18-8**
	16.0	**21-5**	**20-3**	**19-6**	**18-2**	**17-6**	**17-1**	**16-1**
	24.0	**17-5**	**16-7**	**15-11**	**14-10**	**14-3**	**13-11**	**13-4**

*Size = nominal lumber size given in inches
†O.C. = on-center spacing given in inches

TABLE NO. 11. Rafters—High Slope (over 3 in 12)—With No Finished Ceiling. 20 psf live load + 15 psf dead load—heavy roofing. (Spans shown in lightface type are based on a deflection limitation of $l/180$. Spans shown in **boldface type** are limited by allowable extreme fiber stress in bending of the grade and include a 15 psf dead load.)

GRADE

SIZE*	O.C.†	Dense Sel Str KD	Dense Sel Str	Sel Str KD	Sel Str	No. 1 Dense KD	No. 1 Dense	Appearance KD and No. 1 KD	No. 2 Dense KD	No. 1	No. 2 Dense	No. 2 MG KD
2 × 4	12.0	11-8	11-5	11-5	11-3	11-8	11-5	**11-4**	**11-2**	10-10	10-8	**10-3**
	16.0	10-6	10-4	10-4	10-3	10-7	10-1	**9-10**	**9-8**	**9-4**	9-2	**8-10**
	24.0	9-2	11-0	**8-8**	**8-4**	**8-8**	**8-3**	**8-0**	**7-11**	**7-8**	**7-6**	**7-3**
2 × 6	12.0	18-3	18-0	18-0	10-11	17-10	17-2	16-7	16-1	15-10	15-5	14-11
	16.0	16-7	16-1	15-7	9-5	15-5	14-10	14-4	13-11	13-9	13-4	12-11
	24.0	**13-8**	**13-2**	**12-9**	7-9	12-7	12-2	11-8	**11-5**	11-3	10-11	**10-6**
2 × 8	12.0	24-1	23-9	23-9	22-7	23-6	22-7	21-10	21-3	20-11	20-2	19-6
	16.0	21-11	21-3	**20-6**	19-7	20-4	19-7	18-11	18-5	18-1	17-6	16-11
	24.0	**18-0**	**17-4**	**16-9**	**16-0**	**16-7**	**16-0**	**15-5**	**15-0**	14-9	14-4	13-10
2 × 10	12.0	30-6	30-3	30-3	28-11	29-11	28-11	27-10	27-1	26-8	25-11	25-1
	16.0	27-11	27-1	26-2	25-0	25-11	25-0	24-1	23-5	23-1	22-5	21-8
	24.0	**22-1**	**22-1**	**21-5**	**20-3**	**21-2**	**20-5**	**19-8**	**19-2**	18-10	18-3	**17-8**

SIZE*	O.C.†	No. 2 MG	No. 2 KD	No. 2	No. 3 Dense KD	No. 3 Dense	No. 3 KD	No. 3	Construction KD	Construction	Standard	Standard KD
2 × 4	12.0	9-9	9-6	9-2	**8-4**	**8-0**	7-8	7-6	**8-8**	**8-4**	6-7	6-5
	16.0	8-5	8-2	7-11	7-3	**6-11**	6-7	6-5	**7-6**	7-3	5-8	5-6
	24.0	6-11	6-8	6-6	5-11	5-8	5-5	5-3	**6-2**	5-11	4-8	4-6
2 × 6	12.0	14-5	13-8	13-2	12-3	11-10	11-6	10-10				
	16.0	12-6	11-10	11-5	10-8	10-3	9-11	9-6				
	24.0	10-2	9-8	9-4	8-8	8-4	8-2	7-8				
2 × 8	12.0	18-11	18-0	17-4	16-2	15-7	15-2	14-4				
	16.0	16-5	15-7	15-0	14-0	13-6	13-2	12-5				
	24.0	13-5	12-9	12-3	11-5	11-0	10-8	10-1				
2 × 10	12.0	24-3	23-0	22-1	20-7	19-11	19-5	18-4				
	16.0	21-0	19-11	19-2	17-10	17-3	16-9	15-10				
	24.0	16-9	16-3	15-8	14-7	14-0	13-8	12-11				

*Size = nominal lumber size given in inches
†O.C. = on-center spacing given in inches

TABLE NO. 12. Rafters—High Slope (over 3 in 12)—With No Finished Ceiling—30 psf live load + 15 psf dead load—heavy roofing. (Spans shown in lightface type are based on a deflection limitation of $l/180$. Spans shown in **boldface type** are limited by allowable extreme fiber stress in bending of the grade and include a 15 psf dead load.)

GRADE

SIZE*	O.C.†	Dense Sel Str KD	Dense Sel Str	Sel Str KD	Sel Str	No. 1 Dense KD	No. 1 Dense	Appearance KD and No. 1 KD	No. 2 Dense KD	No. 1	No. 2 Dense	No. 2 MG KD
2 × 4	12.0	10-2	10-0	10-0	9-10	10-2	10-2	10-0	9-10	**9-6**	**9-4**	**9-0**
	16.0	9-2	9-0	9-0	8-10	9-2	**8-11**	**8-8**	**8-6**	**8-3**	**8-1**	**7-10**
	24.0	8-1	7-11	7-8	7-5	7-8	7-4	7-1	7-0	**6-9**	**6-8**	**6-5**
2 × 6	12.0	16-0	15-8	15-8	15-2	15-8	15-2	14-7	14-2	14-0	13-8	13-2
	16.0	14-6	14-2	13-8	13-1	13-7	13-1	12-8	12-4	12-1	11-8	11-4
	24.0	12-1	11-7	11-2	10-9	11-1	10-8	10-4	10-0	9-10	9-7	9-3
2 × 8	12.0	21-1	20-8	20-8	20-0	20-8	20-0	19-3	18-9	18-5	17-10	17-3
	16.0	19-2	18-9	18-1	17-3	17-11	17-3	16-8	16-3	16-0	15-6	15-0
	24.0	15-11	15-3	14-10	14-1	14-8	14-1	13-7	13-3	13-0	15-6	14-8
2 × 10	12.0	26-10	26-5	26-5	25-6	26-5	25-6	24-6	23-11	23-7	22-10	22-1
	16.0	24-5	23-11	23-1	22-1	22-10	22-1	21-3	**20-8**	**20-4**	**19-9**	**19-1**
	24.0	20-3	19-6	18-10	18-0	18-8	18-0	17-4	**16-11**	**16-8**	**16-1**	**15-7**

SIZE*	O.C.†	No. 2 MG	No. 2 KD	No. 2	No. 3 Dense KD	No. 3 Dense	No. 3 KD	No. 3	Construction KD	Construction	Standard KD	Standard
2 × 4	12.0	8-7	**8-4**	8-1	7-5	6-11	6-9	6-7	**7-8**	**7-5**	**5-10**	**5-8**
	16.0	**7-6**	**7-3**	**7-0**	6-5	6-0	**5-10**	**5-8**	**6-8**	**6-5**	**5-1**	**4-11**
	24.0	**6-1**	**5-11**	**5-9**	5-3	4-11	4-9	4-7	**5-5**	**5-3**	**4-2**	**4-0**
2 × 6	12.0	**12-9**	**12-1**	**11-7**	10-9	10-5	10-1	9-7				
	16.0	**10-11**	**10-5**	**10-0**	9-4	9-0	8-9	8-3				
	24.0	**8-11**	**8-6**	**8-2**	7-7	7-4	7-2	6-9				
2 × 8	12.0	**16-9**	**15-11**	**15-3**	14-3	13-9	13-5	12-8				
	16.0	**14-6**	**13-9**	**13-3**	12-4	11-11	11-7	11-0				
	24.0	**11-10**	**11-3**	**10-10**	10-1	9-8	9-5	9-0				
2 × 10	12.0	**21-5**	**20-4**	**19-6**	18-2	17-7	17-1	16-2				
	16.0	**18-6**	**17-7**	**16-11**	15-9	15-2	14-10	13-11				
	24.0	**15-1**	**14-4**	**13-9**	12-10	12-5	12-1	11-5				

*Size = nominal lumber size given in inches
†O.C. = on-center spacing given in inches

TABLE NO. 13. Rafters—High Slope (over 3 in 12)—With No Finished Ceiling—20 psf live load + 7 psf dead load—light roofing. (Spans shown in lightface type are based on a deflection limitation of $l/180$. Spans shown in **boldface type** are limited by allowable extreme fiber stress in bending of the grade and include a 7 psf dead load.)

GRADE

SIZE*	O.C.†	Dense Sel Str KD	Dense Sel Str	Sel Str KD	Sel Str	No. 1 Dense KD	No. 1 Dense	Appearance KD and No. 1 KD	No. 2 Dense KD	No. 1	No. 2 Dense	No. 2 MG KD
2 × 4	12.0	11-8	11-5	11-5	11-3	11-8	11-5	11-5	11-3	11-3	11-0	11-0
	16.0	10-9	10-5	10-5	10-3	10-9	10-5	10-5	10-3	10-3	10-0	10-0
	24.0	9-2	9-1	9-1	8-11	9-2	9-1	9-1	8-11	**8-8**	**8-7**	**8-3**
2 × 6	12.0	18-4	18-0	18-0	17-8	18-4	18-0	18-0	17-8	17-8	17-3	**17-0**
	16.0	16-8	16-4	16-4	16-1	16-8	16-4	**16-4**	**15-11**	**15-8**	**15-2**	**14-9**
	24.0	14-6	14-4	14-4	**13-10**	**14-4**	**13-10**	**13-4**	**13-0**	**12-9**	**12-5**	**12-0**
2 × 8	12.0	24-2	23-9	23-9	23-3	24-2	23-9	23-9	23-3	23-1	22-9	**22-5**
	16.0	22-0	21-6	21-6	21-2	22-0	21-6	**21-6**	**20-11**	**20-7**	**20-0**	**19-5**
	24.0	19-2	18-11	18-11	**18-2**	**18-11**	**18-2**	**17-7**	**17-1**	**16-10**	**16-4**	**15-10**
2 × 10	12.0	30-10	30-3	30-3	29-9	30-10	30-3	30-3	29-9	29-5	29-1	**28-7**
	16.0	28-1	27-5	27-5	27-0	28-1	27-5	**27-5**	**26-8**	**26-3**	**25-6**	**24-9**
	24.0	24-5	24-1	24-1	**23-3**	**24-1**	**23-3**	**22-5**	**21-10**	**21-6**	**20-10**	**20-3**

SIZE*	O.C.†	No. 2 MG	No. 2 KD	No. 2	No. 3 Dense KD	No. 3 Dense	No. 3 KD	No. 3	Construction KD	Construction	Standard KD	Standard
2 × 4	12.0	10-10	10-8	10-3	**9-6**	**9-1**	**8-8**	**8-5**	**9-11**	**9-6**	**7-6**	**7-3**
	16.0	**9-8**	**9-4**	**9-1**	**8-3**	**7-11**	**7-6**	**7-4**	**8-7**	**8-3**	**6-6**	**6-4**
	24.0	**7-10**	**7-7**	**7-4**	**6-9**	**6-5**	**6-2**	**6-0**	**7-0**	**6-9**	**5-4**	**5-2**
2 × 6	12.0	**16-5**	**15-7**	**15-0**	**14-0**	**13-6**	**13-2**	**12-5**				
	16.0	**14-3**	**13-6**	**13-0**	**11-1**	**11-8**	**11-4**	**10-8**				
	24.0	**11-7**	**11-0**	**10-7**	**9-11**	**9-6**	**9-3**	**8-9**				
2 × 8	12.0	**21-8**	**20-6**	**19-9**	**18-5**	**17-9**	**17-3**	**16-3**				
	16.0	**18-9**	**17-9**	**17-1**	**15-11**	**15-4**	**15-0**	**14-1**				
	24.0	**15-4**	**14-6**	**13-11**	**13-0**	**12-7**	**12-3**	**11-6**				
2 × 10	12.0	**27-8**	**26-2**	**25-2**	**23-6**	**22-8**	**22-1**	**20-10**				
	16.0	**24-0**	**22-8**	**21-10**	**20-4**	**19-8**	**19-2**	**18-1**				
	24.0	**19-7**	**18-6**	**17-10**	**16-8**	**16-0**	**15-7**	**14-9**				

*Size = nominal lumber size given in inches
†O.C. = on-center spacing given in inches

TABLE NO. 14. Rafters—High Slope (over 3 in 12)—With No Finished Ceiling—30 psf live load + 7 psf dead load—light roofing. (Spans shown in lightface type are based on a deflection limitation of $l/180$. Spans shown in **boldface type** are limited by allowable extreme fiber stress in bending of the grade and include a 7 psf dead load.)

GRADE

SIZE*	O.C.†	Dense Sel Str KD	Dense Sel Str	Sel Str KD	Sel Str	No. 1 Dense KD	No. 1 Dense	Appearance KD and No. 1 KD	No. 2 Dense KD	No. 1	No. 2 Dense	No. 2 MG KD
2 × 4	12.0	10-3	10-0	10-0	9-10	10-3	10-0	10-0	9-10	9-10	9-8	9-8
	16.0	9-3	9-1	9-1	8-11	9-3	9-1	9-1	8-11	8-11	8-9	**8-8**
	24.0	8-0	7-11	7-11	7-9	8-0	7-11	**7-9**	**7-8**	**7-5**	**7-4**	**7-1**
2 × 6	12.0	16-0	15-9	15-9	15-5	16-0	15-9	15-9	15-5	**15-5**	**15-0**	**14-6**
	16.0	14-6	14-4	14-4	14-0	14-6	14-4	**13-11**	**13-7**	**13-4**	**12-11**	**12-7**
	24.0	12-7	12-7	**12-4**	11-10	12-3	11-10	11-5	11-1	**10-11**	**10-7**	**10-3**
2 × 8	12.0	21-2	20-9	20-9	20-4	21-2	20-9	20-9	20-4	**20-4**	**19-9**	**19-2**
	16.0	19-2	18-10	18-10	18-6	19-2	18-10	**18-4**	**17-11**	**17-7**	**17-1**	**16-7**
	24.0	16-9	16-7	**16-4**	15-7	16-2	15-7	15-0	14-7	**14-4**	**13-11**	**13-6**
2 × 10	12.0	27-0	26-5	26-5	25-11	27-0	26-5	26-5	25-11	**25-11**	**25-2**	**24-5**
	16.0	23-9	23-2	23-2	22-10	23-9	23-2	23-4	**22-10**	**22-6**	**21-10**	**21-2**
	24.0	21-3	21-0	**20-10**	19-10	20-7	19-10	19-2	**18-7**	**18-4**	**17-10**	**17-3**

SIZE*	O.C.†	No. 2 MG	No. 2 KD	No. 2	No. 3 Dense KD	No. 3 Dense	No. 3 KD	No. 3	Construction KD	Construction	Standard KD	Standard
2 × 4	12.0	9-5	9-3	**8-11**	**8-2**	**7-9**	**7-5**	**7-3**	**8-6**	**8-2**	**6-5**	**6-3**
	16.0	8-3	8-0	**7-9**	**7-1**	**6-9**	**6-5**	**6-3**	**7-4**	**7-1**	**5-7**	**5-5**
	24.0	6-9	6-6	**6-4**	**5-9**	**5-6**	**5-3**	**5-1**	**6-0**	**5-9**	**4-7**	**4-5**
2 × 6	12.0	**14-1**	**13-4**	**12-9**	**11-11**	**11-6**	**11-2**	**10-7**				
	16.0	**12-2**	**11-6**	**11-1**	**10-4**	**9-11**	**9-8**	**9-2**				
	24.0	**9-11**	**9-5**	**9-1**	**8-5**	**8-1**	**7-11**	**7-6**				
2 × 8	12.0	**18-6**	**17-7**	**16-10**	**15-9**	**15-2**	**14-9**	**13-11**				
	16.0	**16-0**	**15-2**	**14-7**	**13-8**	**13-2**	**12-10**	**12-1**				
	24.0	**13-1**	**12-5**	**11-11**	**11-2**	**10-9**	**10-5**	**9-10**				
2 × 10	12.0	**23-8**	**22-5**	**21-6**	**20-1**	**19-4**	**18-10**	**17-10**				
	16.0	**20-5**	**19-5**	**18-7**	**17-5**	**16-9**	**16-4**	**15-5**				
	24.0	**16-8**	**15-10**	**15-2**	**14-3**	**13-8**	**13-4**	**12-7**				

*Size = nominal lumber size given in inches
†O.C. = on-center spacing given in inches

Plank and Beam Frames

(Courtesy of National Forest Products Association)

DESIGN DATA FOR PLANKS

Design data for plank floors and roofs are included in Table D-1. Computations for bending are based on the live load indicated, plus 10 pounds per square foot of dead load. Computations for deflection are based on the live load only. The table shows four general arrangements of planks as follows:

Type A: Extending over a single span.
Type B: Continuous over two equal spans.
Type C: Continuous over three equal spans.
Type D: A combination of Types A and B.

On the basis of a section of planking 12 inches wide the following formulas were used in making the computations:

For Type A: $\qquad M = \dfrac{wL^2}{8}$ and $D = \dfrac{5wL^4(12)^3}{384EI}$

For Type B: $\qquad M = \dfrac{wL^2}{8}$ and $D = \dfrac{wL^4(12)^3}{185EI}$

For Type C: $\qquad M = \dfrac{wL^2}{10}$ and $D = \dfrac{4wL^4(12)^3}{581EI}$

For Type D: $M = \dfrac{wL^2}{8}$ and $D = \dfrac{1}{2}\left(\dfrac{5wL^4(12)^3}{384EI} + \dfrac{wL^4(12)^3}{185EI}\right)$

To use Table D-1, first determine the plank arrangement (Types A, B, C or D), the span, the live load to be supported and the deflection limitation. Then select from the table the corresponding required values for fiber stress in bending (f) and modulus of elasticity (E). The plank to be used should be of a grade and species which meets these minimum values. The maximum span for a specific grade and species of plank may be determined by reversing these steps.

For those who prefer to use random length planks (instead of arrangements Type A, B, C or D), similar technical information concerning it is included in *Random Length Wood Decking*, a publication of the National Lumber Manufacturers Association.

DESIGN DATA FOR BEAMS

Design data for beams are included in Tables D–2, D–3, and D-4. Computations for bending are based on the live load indicated plus 10 pounds per square foot of dead load. Computations for deflection are based on the live load only. All beams in the tables are designed to extend over a single span and the following formulas are used:

For Type A: $$M = \frac{wL^2}{8} \text{ and } D = \frac{5wL^4 \, (12)^3}{384EI}$$

To use these tables, determine 1) the span, 2) the live load to be supported and, 3) the deflection limitation. Then, from the tables, select the proper size of beam with the corresponding required values for fiber stress in bending (f) and modulus of elasticity (E). The beam used should be of a grade and species which meets these minimum values. The maximum span for a beam of specific size, grade and species can be determined by reversing these steps.

Meanings of Symbols

In the preceding formulas and in the tables the symbols have the following meanings:

w = load in pounds per linear foot.

L = span in feet.

M = induced bending moment in pound-feet.

f = fiber stress in bending in pounds per square inch.

E = modulus of elasticity in pounds per square inch.

I = moment of inertia in inches to the fourth power.

D = deflection in inches.

Lumber Sizes

Tabular data provided herein are based on net dimensions (S4S) as listed in American Softwood Lumber Standard, VPS 20–70.

367

Table D-1 NOMINAL TWO-INCH PLANK

Required values for fiber stress in bending (f) and modulus of elasticity (E) to support safely a live load of 20, 30 or 40 pounds per square foot within a deflection limitation of $l/240$, $l/300$ or $l/360$.

PLANK SPAN IN FEET	LIVE LOAD psf	DEFLECTION LIMITATION	TYPE A		TYPE B		TYPE C		TYPE D	
			f psi	E psi	f psi	E psi	f psi	E psi	f psi	E psi
6'	20	$\frac{l}{240}$	360	576000	360	239000	288	305000	360	408000
		$\frac{l}{300}$	360	720000	360	299000	288	381000	360	509000
		$\frac{l}{360}$	360	864000	360	359000	288	457000	360	611000
	30	$\frac{l}{240}$	480	864000	480	359000	384	457000	480	611000
		$\frac{l}{300}$	480	1080000	480	448000	384	571000	480	764000
		$\frac{l}{360}$	480	1296000	480	538000	384	685000	480	917000
	40	$\frac{l}{240}$	600	1152000	600	478000	480	609000	600	815000
		$\frac{l}{300}$	600	1440000	600	598000	480	762000	600	1019000
		$\frac{l}{360}$	600	1728000	600	717000	480	914000	600	1223000
	20	$\frac{l}{240}$	490	915000	490	380000	392	484000	490	647000
		$\frac{l}{300}$	490	1143000	490	475000	392	605000	490	809000
		$\frac{l}{360}$	490	1372000	490	570000	392	726000	490	971000

7'	30	$\frac{l}{240}$	653	1372000	653	570000	522	726000	653	971000
		$\frac{l}{300}$	653	1715000	653	712000	522	907000	653	1213000
		$\frac{l}{360}$	653	2058000	653	854000	522	1088000	653	1456000
	40	$\frac{l}{240}$	817	1829000	817	759000	653	968000	817	1294000
		$\frac{l}{300}$	817	2287000	817	949000	653	1209000	817	1618000
		$\frac{l}{360}$	817	2744000	817	1139000	653	1451000	817	1941000
8'	20	$\frac{l}{240}$	640	1365000	640	567000	512	722000	640	966000
		$\frac{l}{300}$	640	1707000	640	708000	512	903000	640	1208000
		$\frac{l}{360}$	640	2048000	640	850000	512	1083000	640	1449000
	30	$\frac{l}{240}$	853	2048000	853	850000	682	1083000	853	1449000
		$\frac{l}{300}$	853	2560000	853	1063000	682	1354000	853	1811000
		$\frac{l}{360}$	853	3072000	853	1275000	682	1625000	853	2174000
	40	$\frac{l}{240}$	1067	2731000	1067	1134000	853	1444000	1067	1932000
		$\frac{l}{300}$	1067	3413000	1067	1417000	853	1805000	1067	2415000
		$\frac{l}{360}$	1067	4096000	1067	1700000	853	2166000	1067	2898000

Table D-2 FLOOR AND ROOF BEAMS

Required values for fiber stress in bending (f) and modulus of elasticity (E) for the sizes shown to support safely a live load of 20 pounds per square foot within a deflection limitation of $l/360$.

SPAN OF BEAM	NOMINAL SIZE OF BEAM	6'-0" f	6'-0" E	7'-0" f	7'-0" E	8'-0" f	8'-0" E
10'	2-3x6	1020	1110000	1190	1300000	1360	1480000
	1-3x8	1100	875000	1280	1020000	1460	1170000
	2-2x8	890	710000	1030	825000	1180	940000
	1-4x8	790	635000	920	740000	1050	845000
	3-2x8	590	475000	690	550000	790	630000
	2-3x8	550	440000	640	510000	730	585000
	1-6x8	520	520000	610	490000	700	560000
11'	2-3x6	1230	1480000	1440	1730000		
	1-3x8	1320	1170000	1540	1360000	1760	1550000
	2-2x8	1070	945000	1250	1100000	1430	1260000
	1-4x8	960	845000	1120	985000	1270	1130000
	3-2x8	710	630000	830	735000	950	840000
	2-3x8	660	585000	780	680000	880	775000
	1-6x8	630	560000	740	650000	840	745000
12'	1-3x8	1580	1510000	1840	1760000		
	2-2x8	1270	1220000	1490	1430000	1700	1640000
	1-4x8	1140	1100000	1330	1280000	1510	1460000
	3-2x8	850	815000	990	955000	1130	1090000
	2-3x8	790	755000	920	882000	1050	1010000
	1-6x8	750	720000	880	845000	1000	965000
	2-2x10	790	600000	920	705000	1060	805000
	1-3x10	980	745000	1150	870000	1310	995000
13'	2-2x8	1500	1560000	1740	1820000		
	1-4x8	1340	1400000	1550	1630000		
	3-2x8	990	1040000	1160	1220000	1330	1400000
	2-3x8	920	965000	1080	1130000	1230	1290000
	1-6x8	880	925000	1030	1080000	1180	1240000
	2-2x10	930	770000	1080	900000	1240	1030000
	1-3x10	1150	950000	1350	1110000	1540	1270000
	1-4x10	830	690000	970	805000	1110	920000
14'	3-2x8	1150	1300000	1350	1510000	1540	1730000
	2-3x8	1070	1200000	1250	1400000	1430	1600000
	1-6x8	1030	1150000	1200	1340000	1370	1530000
	1-3x10	1340	1180000	1570	1380000	1780	1580000
	2-2x10	1080	960000	1260	1120000	1440	1280000
	1-4x10	970	860000	1130	1000000	1290	1140000
	3-2x10	720	640000	840	745000	960	855000
	2-3x10	670	590000	780	690000	890	790000
	1-6x10	640	565000	750	660000	850	755000
	4-2x10	540	480000	630	560000	720	640000
	2-2x12	740	540000	870	630000	990	720000
15'	3-2x8	1320	1590000	1440	1720000		
	2-3x8	1230	1480000				
	1-6x8	1180	1410000	1370	1650000		
	1-3x10	1540	1460000	1800	1700000		
	2-2x10	1240	1180000	1450	1380000	1650	1570000
	1-4x10	1110	1050000	1300	1230000	1480	1400000
	3-2x10	820	786000	960	918000	1100	1050000
	2-3x10	770	730000	900	850000	1030	970000
	1-6x10	740	695000	860	810000	980	925000
	4-2x10	620	590000	720	690000	830	785000
	2-2x12	850	665000	990	775000	1140	885000
	1-4x12	760	594000	890	695000	1020	792000
16'	2-2x10	1410	1440000	1650	1670000		
	1-4x10	1260	1280000	1470	1490000	1680	1710000
	3-2x10	940	960000	1100	1110000	1250	1270000
	2-3x10	880	886000	1020	1030000	1170	1180000
	1-6x10	840	845000	970	985000	1110	1130000
	4-2x10	700	720000	820	835000	940	955000
	1-8x10	610	620000	710	720000	810	830000
	1-3x12	1200	1000000	1400	1177000	1600	1330000
	2-2x12	970	810000	1130	940000	1290	1080000
	1-4x12	870	725000	1010	845000	1150	965000
	3-2x12	650	540000	750	630000	860	720000
	2-3x12	600	500000	700	588000	800	665000

SPAN OF BEAM	NOMINAL SIZE OF BEAM	6'-0" f	6'-0" E	7'-0" f	7'-0" E	8'-0" f	8'-0" E
17'	2-2x10	1590	1720000				
	1-4x10	1430	1530000	1670	1790000		
	3-2x10	1060	1140000	1240	1330000	1410	1530000
	2-3x10	990	1060000	1150	1240000	1310	1410000
	1-6x10	940	1010000	1100	1180000	1250	1350000
	4-2x10	790	860000	930	1000000	1060	1140000
	1-8x10	690	740000	810	865000	920	990000
	1-3x12	1350	1200000	1580	1400000	1800	1600000
	2-2x12	1090	970000	1280	1130000	1450	1290000
	1-4x12	980	865000	1140	1010000	1300	1150000
	3-2x12	730	645000	850	740000	970	858000
	2-3x12	680	600000	790	700000	900	800000
18'	3-2x10	1190	1360000	1370	1590000		
	2-3x10	1110	1260000	1300	1470000	1480	1680000
	1-6x10	1060	1200000	1230	1400000	1410	1600000
	4-2x10	890	1020000	1040	1190000	1190	1360000
	1-8x10	780	840000	910	1030000	1030	1180000
	1-3x12	1520	1420000	1730	1660000		
	2-2x12	1230	1150000	1430	1340000	1630	1530000
	1-4x12	1100	1030000	1280	1200000	1460	1370000
	3-2x12	820	765000	960	895000	1090	1020000
	2-3x12	760	710000	890	830000	1010	950000
	1-6x12	720	680000	840	790000	960	900000
	4-2x12	620	575000	720	670000	820	765000
19'	3-2x10	1320	1600000				
	2-3x10	1230	1480000	1440	1730000		
	1-6x10	1170	1410000	1370	1650000		
	4-2x10	990	1200000	1160	1400000	1320	1600000
	2-4x10	890	1070000	1040	1250000	1190	1430000
	1-8x10	860	1040000	1010	1210000	1150	1380000
	2-2x12	1360	1350000	1590	1580000		
	1-4x12	1220	1210000	1430	1410000	1630	1610000
	3-2x12	910	900000	1060	1050000	1210	1200000
	2-3x12	840	835000	980	980000	1130	1110000
	1-6x12	800	800000	940	930000	1070	1060000
	4-2x12	680	675000	790	790000	910	900000
20'	2-3x10	1370	1730000				
	1-6x10	1300	1650000				
	4-2x10	1100	1400000	1280	1630000		
	2-4x10	990	1250000	1150	1460000	1320	1660000
	1-8x10	960	1210000	1120	1410000	1280	1610000
	2-2x12	1510	1580000				
	1-4x12	1350	1410000	1580	1640000		
	3-2x12	1010	1050000	1180	1220000	1350	1400000
	2-3x12	940	975000	1090	1140000	1250	1300000
	1-6x12	890	930000	1040	1090000	1190	1240000
	4-2x12	760	790000	880	920000	1010	1050000
	2-4x12	680	705000	790	820000	900	940000
21'	4-2x10	1210	1620000				
	2-4x10	1090	1450000	1270	1690000		
	1-8x10	1050	1400000	1230	1630000		
	1-4x12	1490	1630000				
	3-2x12	1110	1210000	1300	1420000	1480	1620000
	2-3x12	1030	1130000	1200	1320000	1380	1500000
	1-6x12	970	1070000	1130	1250000	1310	1430000
	4-2x12	830	910000	970	1060000	1110	1210000
	2-4x12	750	815000	870	950000	990	1090000
	3-3x12	690	750000	800	875000	920	1000000
	1-8x12	720	790000	840	920000	960	1050000
	1-10x12	580	620000	660	725000	750	826000
22'	2-4x10	1200	1670000				
	1-8x10	1160	1610000				
	3-2x12	1220	1400000	1420	1630000		
	2-3x12	1130	1300000	1320	1510000	1520	1730000
	1-6x12	1080	1240000	1250	1450000	1450	1650000
	4-2x12	920	1050000	1070	1230000	1230	1400000
	2-4x12	820	940000	960	1090000	1100	1250000
	5-2x12	730	840000	860	980000	980	1120000
	3-3x12	750	865000	880	1010000	1010	1160000
	1-8x12	790	910000	920	1060000	1060	1210000
	1-10x12	620	715000	730	835000	830	955000

22

Table D-3 FLOOR AND ROOF BEAMS

Required values for fiber stress in bending (f) and modulus of elasticity (E) for the sizes shown to support safely a live load of 30 pounds per square foot within a deflection limitation of $l/360$.

SPAN OF BEAM	NOMINAL SIZE OF BEAM	6'-0" f	6'-0" E	7'-0" f	7'-0" E	8'-0" f	8'-0" E
10'	2-3x6	1360	1670000				
	1-3x8	1460	1320000	1700	1540000	1950	1740000
	1-4x8	1050	955000	1220	1120000	1400	1275000
	3-2x8	790	710000	920	830000	1050	950000
	2-3x8	730	655000	850	770000	970	880000
	2-4x8	530	475000	610	560000	700	635000
	2-2x10	740	525000	860	615000	980	705000
11'	1-3x8	1770	1750000				
	1-4x8	1270	1270000	1490	1480000	1700	1690000
	3-2x8	950	950000	1110	1100000	1270	1260000
	2-3x8	880	875000	1030	1020000	1180	1170000
	2-4x8	640	635000	750	740000	850	845000
	2-2x10	890	700000	1040	815000	1180	930000
	1-3x10	1100	865000	1290	1010000	1480	1150000
12'	1-4x8	1510	1650000				
	3-2x8	1130	1230000	1320	1430000	1510	1640000
	2-3x8	1050	1040000	1230	1320000	1400	1510000
	2-4x8	760	825000	880	865000	1010	1100000
	1-6x8	1000	1090000	1170	1270000	1340	1450000
	2-2x10	1060	910000	1230	1060000	1410	1210000
	3-2x10	700	605000	820	705000	940	805000
	2-3x10	660	560000	760	655000	880	750000
	1-4x10	950	810000	1110	945000	1270	1080000
13'	3-2x8	1330	1560000				
	2-3x8	1230	1440000	1440	1690000		
	2-4x8	890	1050000	1040	1220000	1190	1390000
	1-6x8	1180	1380000	1370	1610000		
	2-2x10	1240	1150000	1450	1350000	1650	1540000
	3-2x10	830	835000	970	900000	1100	1030000
	2-3x10	770	710000	900	830000	1030	950000
	1-4x10	1110	1030000	1300	1080000	1480	1370000
	1-6x10	730	680000	860	795000	980	905000
14'	2-4x8	1030	1310000	1210	1520000	1380	1740000
	1-6x8	1370	1720000				
	2-2x10	1490	1440000	1680	1680000		
	3-2x10	960	960000	1120	1120000	1280	1280000
	2-3x10	890	890000	1050	1040000	1200	1190000
	1-4x10	1290	1280000	1510	1500000	1720	1720000
	1-6x10	850	850000	1000	990000	1140	1130000
	2-4x10	650	645000	750	750000	860	860000
	4-2x10	720	720000	840	840000	960	960000
	1-8x10	630	620000	730	725000	840	830000
15'	2-2x10	1650	1770000				
	3-2x10	1100	1180000	1280	1380000	1470	1570000
	2-3x10	1030	1090000	1200	1280000	1370	1460000
	1-4x10	1480	1580000				
	1-6x10	980	1040000	1140	1220000	1300	1390000
	2-4x10	740	790000	865	925000	990	1060000
	4-2x10	830	885000	960	1030000	1100	1180000
	1-8x10	720	765000	840	890000	960	1020000
	2-2x12	1040	995000	1210	1160000	1390	1330000
	1-4x12	1020	890000	1190	1040000	1350	1190000
	3-2x12	760	665000	880	775000	1000	885000
	2-3x12	700	615000	820	720000	940	820000
16'	3-2x10	1250	1430000	1460	1670000		
	2-3x10	1170	1330000	1360	1550000	1550	1770000
	1-6x10	1110	1270000	1300	1450000	1480	1680000
	2-4x10	840	960000	980	1120000	1120	1280000
	4-2x10	940	1070000	1100	1250000	1250	1430000
	1-8x10	830	930000	950	1080000	1090	1240000
	2-2x12	1290	1210000	1510	1410000	1720	1610000
	1-4x12	1160	1080000	1350	1260000	1540	1450000
	3-2x12	860	805000	1000	940000	1150	1070000
	2-3x12	800	750000	930	875000	1070	1000000
	4-2x12	640	605000	750	700000	860	795000
	2-4x12	580	545000	670	635000	770	725000

SPAN OF BEAM	NOMINAL SIZE OF BEAM	6'-0" f	6'-0" E	7'-0" f	7'-0" E	8'-0" f	8'-0" E
17'	2-3x10	1320	1590000				
	1-6x10	1260	1520000	1470	1770000		
	2-4x10	950	1150000	1110	1340000	1270	1540000
	4-2x10	1060	1290000	1240	1500000	1420	1720000
	1-8x10	920	1110000	1080	1300000	1230	1490000
	2-2x12	1460	1450000	1700	1690000		
	1-4x12	1310	1300000	1520	1520000	1750	1730000
	3-2x12	970	965000	1140	1130000	1300	1290000
	2-3x12	900	900000	1060	1050000	1210	1200000
	4-2x12	730	725000	850	845000	970	965000
	2-4x12	650	650000	760	760000	870	870000
	5-2x12	580	580000	680	675000	780	775000
18'	2-4x10	1070	1360000	1250	1590000		
	4-2x10	1190	1530000	1390	1780000		
	1-8x10	1040	1320000	1210	1540000	1380	1760000
	2-2x12	1640	1760000				
	1-4x12	1470	1540000				
	3-2x12	1090	1140000	1270	1340000	1460	1530000
	2-3x12	1010	1070000	1180	1240000	1350	1420000
	4-2x12	820	860000	960	1000000	1090	1150000
	2-4x12	730	770000	860	900000	980	1030000
	5-2x12	660	690000	770	800000	870	920000
	1-6x12	960	1010000	1120	1180000	1280	1350000
	3-3x12	680	710000	790	825000	900	945000
19'	2-4x10	1190	1610000				
	1-8x10	1150	1560000				
	3-2x12	1220	1350000	1420	1580000		
	2-3x12	1140	1260000	1320	1460000	1500	1670000
	4-2x12	910	1010000	1060	1180000	1210	1350000
	2-4x12	820	910000	950	1060000	1090	1210000
	5-2x12	730	810000	850	950000	970	1080000
	1-6x12	1070	1200000	1250	1400000	1430	1600000
	3-3x12	750	835000	880	975000	1000	1110000
	1-8x12	790	880000	920	1030000	1050	1170000
	1-10x12	620	690000	720	810000	830	920000
	4-3x12	563	630000	658	735000	750	835000
20'	3-2x12	1210	1570000				
	2-3x12	1120	1460000	1310	1710000		
	4-2x12	910	1180000	1060	1380000	1210	1570000
	2-4x12	810	1060000	950	1240000	1090	1410000
	5-2x12	730	945000	850	1100000	970	1260000
	1-6x12	1070	1390000	1250	1620000	1430	1850000
	3-3x12	750	970000	880	1130000	1000	1300000
	1-8x12	790	1020000	920	1190000	1050	1370000
	1-10x12	620	805000	720	940000	830	1070000
	4-3x12	560	730000	660	850000	750	975000
	2-3x14	810	905000	950	1050000	1090	1210000
	1-6x14	780	865000	910	1010000	1040	1150000
21'	2-3x12	1380	1760000				
	4-2x12	1110	1420000	1300	1660000		
	2-4x12	1000	1280000	1160	1490000	1330	1700000
	5-2x12	890	1140000	1040	1330000	1190	1520000
	1-6x12	1310	1680000				
	3-3x12	920	1170000	1070	1370000	1220	1560000
	1-8x12	970	1230000	1130	1440000	1290	1640000
	1-10x12	760	970000	880	1130000	1010	1290000
	4-3x12	690	880000	800	1030000	920	1170000
	2-3x14	1000	1090000	1170	1270000	1330	1450000
	1-6x14	950	1040000	1110	1210000	1270	1380000
	3-3x14	670	730000	770	850000	880	970000
22'	4-2x12	1220	1570000				
	2-4x12	1090	1410000	1280	1650000		
	5-2x12	980	1260000	1140	1470000	1300	1680000
	3-3x12	1010	1290000	1180	1510000	1340	1720000
	1-8x12	1050	1360000	1230	1590000		
	1-10x12	830	1070000	970	1250000	1110	1430000
	4-3x12	750	970000	880	1130000	1010	1300000
	2-3x14	1090	1200000	1280	1400000	1460	1610000
	1-6x14	1040	1150000	1220	1340000	1390	1540000
	3-3x14	730	800000	850	935000	970	1070000
	2-4x14	790	870000	920	1010000	1060	1160000
	4-3x14	550	600000	640	700000	730	800000

25

Table D-4 FLOOR AND ROOF BEAMS

Required values for fiber stress in bending (f) and modulus of elasticity (E) for the sizes shown to support safely a live load of 40 pounds per square foot within a deflection limitation of $l/360$.

SPAN OF BEAM	NOMINAL SIZE OF BEAM	6'-0" f	6'-0" E	7'-0" f	7'-0" E	8'-0" f	8'-0" E
10'	1-3x8	1820	1750000				
	2-2x8	1480	1420000	1720	1660000		
	1-4x8	1320	1270000	1540	1480000	1750	1700000
	1-6x8	870	840000	1020	980000	1160	1120000
	2-2x10	920	700000	1070	815000	1220	935000
	1-3x10	1140	865000	1330	1010000	1520	1150000
	1-4x10	820	625000	960	730000	1090	835000
11'	1-4x8	1600	1700000				
	1-6x8	1060	1120000	1230	1300000	1410	1490000
	2-2x10	1110	935000	1300	1080000	1490	1240000
	1-3x10	1380	1150000	1610	1340000	1840	1540000
	1-4x10	1000	835000	:160	970000	1330	1110000
	3-2x10	740	620000	860	725000	990	830000
	2-3x10	690	575000	800	670000	920	770000
12'	1-6x8	1250	1460000	1460	1700000		
	3-2x8	1410	1650000				
	2-2x10	1360	1220000	1560	1420000	1760	1620000
	1-3x10	1630	1500000	1910	1750000		
	1-4x10	1180	1090000	1380	1270000	1580	1450000
	2-3x10	810	750000	950	875000	1090	1000000
	3-2x10	880	810000	1030	945000	1170	1080000
	1-6x10	780	715000	910	835000	1040	955000
	2-4x10	590	545000	690	635000	790	725000
13'	2-4x8	1110	1390000	1300	1630000		
	2-2x10	1550	1530000	1810	1790000		
	3-2x10	1010	1020000	1180	1190000	1380	1370000
	2-3x10	960	950000	1120	1110000	1280	1270000
	1-4x10	1390	1370000	1620	1600000		
	2-4x10	690	685000	810	800000	920	915000
	3-3x10	640	630000	750	740000	860	845000
	1-6x10	920	905000	1070	1050000	1220	1210000
	4-2x10	780	770000	905	895000	1030	1020000
14'	2-4x8	1290	1300000	1510	1520000	1720	1730000
	3-2x10	1200	1280000	1400	1490000	1600	1700000
	2-3x10	1110	1190000	1300	1380000	1490	1580000
	2-4x10	800	860000	940	1000000	1070	1140000
	3-3x10	740	790000	870	920000	990	1050000
	1-6x10	1060	1130000	1240	1320000	1420	1510000
	1-8x10	780	830000	910	970000	1040	1110000
	4-2x10	900	960000	1050	1020000	1200	1280000
	2-2x12	1230	1080000	1440	1260000	1650	1440000
	3-2x12	820	720000	960	840000	1100	960000
15'	3-2x10	1380	1580000				
	2-3x10	1290	1460000	1500	1700000		
	2-4x10	930	1060000	1080	1230000	1240	1410000
	3-3x10	860	975000	1000	1140000	1140	1300000
	1-6x10	1220	1390000	1430	1620000	1630	1850000
	1-8x10	900	1020000	1050	1190000	1200	1360000
	4-2x10	1030	1180000	1210	1380000	1380	1570000
	2-2x12	1420	1330000	1660	1550000	1900	1770000
	3-2x12	950	885000	1110	1030000	1260	1180000
	1-3x12	1760	1650000				
	4-2x12	710	665000	830	775000	950	885000
	2-3x12	880	820000	1030	960000	1170	1100000
16'	2-3x10	1460	1770000				
	2-4x10	1050	1290000	1230	1490000	1410	1710000
	3-3x10	970	1180000	1130	1380000	1290	1570000
	1-6x10	1390	1690000				
	1-8x10	1020	1240000	1190	1440000	1360	1650000
	4-2x10	1170	1430000	1370	1670000		
	2-2x12	1610	1610000				
	3-2x12	1080	1070000	1260	1250000	1440	1430000
	4-2x12	810	806000	940	940000	1080	1070000
	5-2x12	640	645000	750	750000	860	860000
	2-3x12	1000	995000	:170	1160000	1330	1330000
	3-3x12	670	665000	780	775000	890	885000

SPAN OF BEAM	NOMINAL SIZE OF BEAM	6'-0" f	6'-0" E	7'-0" f	7'-0" E	8'-0" f	8'-0" E
17'	2-4x10	1190	1540000	1390	1780000		
	3-3x10	1100	1420000	1280	1640000		
	1-8x10	1150	1490000	1340	1730000		
	3-2x12	1220	1290000	1420	1500000	1620	1710000
	4-2x12	910	1070000	1060	1120000	1210	1280000
	5-2x12	730	775000	850	900000	970	1030000
	2-3x12	1130	1200000	1310	1390000	1500	1590000
	3-3x12	750	795000	880	925000	1000	1060000
	2-4x12	820	870000	950	1010000	1090	1160000
	1-6x12	1070	1140000	1250	1330000	1430	1520000
	1-8x12	790	840000	920	975000	1050	1110000
	3-4x12	540	580000	630	670000	730	720000
18'	3-3x10	1230	1680000				
	1-8x10	1290	1760000				
	3-2x12	1360	1530000	1590	1780000		
	4-2x12	1020	1150000	1190	1340000	1360	1530000
	5-2x12	820	920000	950	1070000	1090	1220000
	2-3x12	1270	1420000	1480	1660000	1680	1890000
	3-3x12	840	945000	980	1100000	1120	1260000
	2-4x12	920	1030000	1070	1200000	1220	1370000
	1-6x12	1200	1350000	1400	1580000	1600	1800000
	1-8x12	880	995000	1030	1160000	1180	1325000
	3-4x12	610	685000	710	800000	810	915000
	4-3x12	630	710000	740	830000	840	945000
19'	4-2x12	1140	1350000	1330	1570000		
	5-2x12	910	1080000	1060	1260000	1210	1440000
	2-3x12	1410	1680000				
	3-3x12	940	1110000	1100	1300000	1250	1480000
	2-4x12	1020	1210000	1190	1410000	1360	1620000
	1-6x12	1340	1590000				
	1-8x12	990	1170000	1150	1360000	1310	1560000
	3-4x12	680	810000	790	940000	910	1070000
	4-3x12	700	835000	820	975000	940	1110000
	2-6x12	670	795000	780	930000	890	1060000
	1-10x12	770	920000	900	1070000	1030	1230000
	2-3x14	1020	1030000	1190	1210000	1360	1380000
20'	4-2x12	1260	1570000				
	5-2x12	1010	1260000	1180	1470000	1350	1680000
	3-3x12	1040	1300000	1210	1510000	1390	1730000
	2-4x12	1130	1420000	1320	1650000		
	1-8x12	1090	1370000	1270	1590000		
	3-4x12	750	945000	880	1100000	1000	1260000
	4-3x12	780	975000	910	1140000	1040	1300000
	2-6x12	740	930000	870	1080000	990	1240000
	1-10x12	850	1070000	1000	1250000	1140	1430000
	2-3x14	1130	1210000	1320	1410000	1510	1610000
	1-6x14	1080	1150000	1260	1340000	1440	1540000
	2-4x14	820	875000	960	1020000	1090	1160000
21'	5-2x12	1110	1450000	1300	1700000		
	3-3x12	1150	1500000	1340	1750000		
	2-4x12	1240	1640000				
	1-8x12	1200	1580000				
	3-4x12	830	1090000	970	1270000	1110	1450000
	4-3x12	860	1130000	1000	1320000	1150	1500000
	2-6x12	820	1070000	960	1260000	1090	1430000
	1-10x12	940	1240000	1100	1450000	1260	1660000
	2-3x14	1240	1400000	1450	1630000		
	1-6x14	1190	1330000	1390	1560000	1590	1780000
	2-4x14	900	1010000	1050	1180000	1200	1350000
	3-3x14	830	930000	970	1090000	1110	1240000
22'	5-2x12	1220	1670000				
	3-3x12	1260	1730000				
	3-4x12	910	1250000	1060	1460000	1210	1670000
	4-3x12	940	1300000	1100	1520000	1260	1730000
	2-6x12	900	1240000	1050	1450000	1200	1650000
	1-10x12	1040	1430000	1210	1670000		
	2-3x14	1360	1610000				
	1-6x14	1310	1530000	1530	1790000		
	2-4x14	990	1160000	1160	1360000	1320	1550000
	3-3x14	910	1070000	1060	1250000	1210	1430000
	3-4x14	660	775000	770	905000	880	1030000
	1-8x14	950	1130000	1100	1310000	1270	1500000

28

372

Reference List

Audel's Carpenter and Builder Guide, Theo. Audel and Co., Publisher, New York, NY.

Badzinski, Stanley, *Carpentry in Residential Construction*, Reston Publishing Co., Reston, VA.

"Grading Rules for Western Lumber," Western Wood Products Association, Portland, OR.

"Guides to Improved Framed Walls for Houses, US," U.S. Department of Agriculture, Madison, WI.

Maguire, Byron W. *Carpentry for Residential Construction*, Reston Publishing Co., Reston, VA.

Maguire, Byron W. *Carpentry in Commercial Construction*, Reston Publishing Co., Reston, VA.

"Manual for House Framing," National Forest Products Association, Washington, D.C.

"Maximum Spans for Joists and Rafters," Southern Forest Products Association, New Orleans, LA.

"MOD 24 Building Guide," American Plywood Association, Tacoma, WA.

"Plank and Beam Framing for Residential Building," National Forest Products Association, Washington, D.C.

"Plywood Residential Construction Guide," American Plywood Association, Tacoma, WA.

"Series 1, Construction Cost Series, Wood Framing Details," National Forest Products Association, Washington, D.C.

"Southern Pine Grade Use Guide," Southern Forest Products Association, New Orleans, LA.

"Span Tables for Joists and Rafters," National Forest Products Association, Washington, D.C.

"Thermax Sheathing System," Celatex Corp., Tampa, FL.

"Western Woods Species Book, Vol. 1." Western Wood Products Association, Portland, OR.

"Wood-Frame House Construction," U.S. Dept. of Agriculture, Handbook No. 73, Washington, D.C.

"Wood Decay in Homes," U.S. Department of Agriculture, Bulletin No. 73, Washington, D.C.

"Wood Preservation," Southern Forest Products Association, New Orleans, LA.

"Working Stresses for Joists and Rafters," National Forest Products Asssociation, Washington, D.C.

"Uniform Building Code," Whittier, CA.

"Zonolite Wall Insulation System," W. R. Grace and Co., Cambridge, MA.

Index